SCIENCE AND BELIEFS

The years between 1700 and 1900 witnessed a fundamental transition in attitudes towards science, as earlier concepts of natural philosophy were replaced with a more modern conception of science. This process was by no means a simple progression, and the changing attitudes to science were marked by bitter arguments and fundamental differences of opinion, many of which are still not entirely resolved today.

Approaching the subject from a number of cultural angles, the essays in this volume explore the fluid relationship between science and belief during this crucial period, and help to trace the development of science as an independent field of study that did not look to religion to provide answers to the workings of the universe.

Taking a broadly chronological approach, each essay in this book addresses a theme that helps illuminate these concerns and highlights how beliefs – both religious and secular – have impinged and influenced the scientific world. By addressing key issues such as the ongoing debate between Christian fundamentalists and followers of Darwin, and the rise of 'respectable atheism', fascinating insights are provided that help to chart the ever-shifting discourse of science and beliefs.

About the Editors

David M. Knight is at the Department of Philosophy at the University of Durham, UK.

Matthew D. Eddy is Lecturer in the History and Philosophy of Science at the University of Durham, UK.

Science, Technology and Culture, 1700–1945

Series Editors

David M. Knight
University of Durham

and

Trevor Levere
University of Toronto

Science, Technology and Culture, 1700–1945 focuses on the social, cultural, industrial and economic contexts of science and technology from the 'scientific revolution' up to the Second World War. It explores the agricultural and industrial revolutions of the eighteenth century, the coffee-house culture of the Enlightenment, the spread of museums, botanic gardens and expositions in the nineteenth century, to the Franco-Prussian war of 1870, seen as a victory for German science. It also addresses the dependence of society on science and technology in the twentieth century.

Science, Technology and Culture, 1700–1945 addresses issues of the interaction of science, technology and culture in the period from 1700 to 1945, at the same time as including new research within the field of the history of science.

Also in this series

Discovering Water
James Watt, Henry Cavendish and the Nineteenth-Century 'Water Controversy'
David Philip Miller

Science and Dissent in England, 1688–1945
Edited by Paul Wood

The Genius of Erasmus Darwin
Edited by C.U.M. Smith and Robert Arnott

John Phillips and the Business of Victorian Science
Jack Morrell

Science and Beliefs

From Natural Philosophy to Natural Science, 1700–1900

Edited by

DAVID M. KNIGHT
University of Durham

MATTHEW D. EDDY
University of Durham

Routledge
Taylor & Francis Group
LONDON AND NEW YORK

First published 2005 by Ashgate Publishing

2 Park Square, Milton Park, Abingdon, Oxon OX14 4RN
711 Third Avenue, New York, NY 10017, USA

Routledge is an imprint of the Taylor & Francis Group, an informa business

First issued in paperback 2017

British Library Cataloguing in Publication Data
Science and beliefs : from natural philosophy to natural
science, 1700–1900. – (Science, technology and culture, 1700–1945)
1. Science – Great Britain – History – 19th century 2. Science – Great Britain
– History – 18th century 3. Science – Social aspects – Great Britain – History –
18th century 4. Science – Social aspects – Great Britain – History – 19th century
5. Religion and science – Great Britain – History – 19th century
I. Knight, David, 1936 Nov. 30– II. Eddy, M.D.
509.4′1′09034

Library of Congress Cataloging in Publication Data
Science and beliefs : from natural philosophy to natural science, 1700–1900 / edited
by David M. Knight and Matthew D. Eddy.
p. cm. – (Science, technology and culture, 1700–1945)
Includes index.
ISBN 0–7546–3996–7 (alk. paper)
1. Science – History. 2. Science and civilization.
I. Knight, David M. II. Eddy, Matthew, 1972– III. Series.

Q125.S1825 2004
509′.033–dc22 2004008415

ISBN 978-0-7546-3996-1 (hbk)
ISBN 978-1-138-25890-7 (pbk)

Contents

PART II: BELIEFS UNDERLYING SCIENCE

CONCLUSION

Acknowledgements

Most of the chapters in this book were originally given as papers at a conference held in 2002 at St John's College, University of Durham, England. Entitled *Science and Beliefs*, the meeting was made possible by funds provided by the Ian Ramsey Study Centre (Oxford), the Center for Theology and the Natural Sciences (Berkeley, CA), the Wellcome Trust and the University of Durham Department of Philosophy. A very special thanks to those who attended the conference and commented on the papers, and to Athanasia Hadjifotiou, Andreas Pantazatos and Mitch Hawkins for their help with planning the event.

List of Contributors

Peter J. Bowler is professor of the History of Science at Queen's University, Belfast. He has written widely on the history of evolution theory and related topics. Among his works are *Reconciling Science and Religion: the Debate in early-twentieth-century Britain* (2002) and *Evolution: the History of an idea* (revised edition, 2003). He is currently working on the involvement of scientists in popular science writing in early twentieth-century Britain.

William H. Brock read Chemistry at University College, London, and then taught History of Science at the University of Leicester between 1960 and 1998. The author of several books on the history of chemistry, including the *Fontana History of Chemistry* (1992), he is currently writing a biography of Sir William Crookes while a visiting professor at the University of Kent at Canterbury.

John H. Brooke is Andreas Idreos professor of Science and Religion, and director of the Ian Ramsey Centre at the University of Oxford where he is also a fellow of Harris Manchester College. His books include *Science and Religion: Some historical perspectives* (1991), which won the Watson Davis Prize of the History of Science Society, and *Reconstructing Nature: the Engagement of science and religion* (1998), jointly authored with Geoffrey Cantor, with whom he gave the Glasgow Gifford Lectures in 1995. He currently directs the European Science Foundation network on 'science and human values'.

Geoffrey Cantor is professor of the History of Science at the University of Leeds and researches the interrelations between science and religion during the eighteenth and nineteenth centuries. Among his publications are *Michael Faraday, Sandemanian and Scientist* (1991) and, with John Hedley Brooke, *Reconstructing Nature: the Engagement of science and religion* (1998).

Randolph Cock is a research fellow in the Centre for Maritime Historical Studies at the University of Exeter, working on a guide to the naval records in the Public Record Office. His research interests centre on naval history, especially in relation to science and exploration.

Thomas Dixon is Lecturer in History at the University of Lancaster and was recently a British Academy postdoctoral fellow at the Faculty of Divinity and a fellow of

Churchill College, University of Cambridge. His first book was *From Passions to Emotions: the Creation of a secular psychological category* (2003).

Matthew D. Eddy is lecturer in the History and Philosophy of Science at the University of Durham. He has most recently held fellowships at the Dibner Institute (MIT), Harvard University, the Max Planck Institute for the History of Science (Berlin) and with the University of Notre Dame's Erasmus Institute. He is currently co-editing the first edition of William Paley's *Natural Theology* (with David M. Knight) and is writing a book on the interaction between eighteenth-century medicine, philosophy and geology.

Aileen Fyfe is lecturer in the History of Science and Medicine at the National University of Ireland, Galway. She works on the popularisation of the sciences to non-expert audiences and on the interaction between religious faith and the sciences. She is the author of *Science and Salvation: Evangelicals and popular science publishing in Victorian Britain* (2004).

Barry Gower is Emeritus Professor of the Philosophy of Science at the University of Durham. He is the author of *Scientific Method: an Historical and philosophical introduction* (1997) and he has published research papers in probabilistic reasoning in science, in relations between metaphysics and science, and in the early twentieth-century origins of analytic philosophy.

Michael Honeybone is an Open University research fellow. His doctoral research was on the dissemination of science among learned societies in the English East Midlands in the early eighteenth century. He is currently editing the correspondence of Maurice Johnson, founder of the Spalding Gentlemen's Society.

Frank A.J.L. James is reader in the History of Science at the Royal Institution, London. His main research concentrates on the physical sciences in the nineteenth century. He is editor of Faraday's correspondence and recently edited *The Common Purposes of Life* (2002) a collection of essays on the Royal Institution.

David M. Knight is Emeritus Professor of the History and Philosophy of Science at the University of Durham. He has written and edited many books, including *Atoms and Elements* (1967), *Natural Science Books* (1972), *The Age of Science* (1986), *Humphry Davy* (1992) and *Science and Spirituality* (2003). His course in Science and Religion received two Templeton Awards and he was recently given the Sidney M. Edelstein Award by the American Chemical Society. He is currently working on a book about the public understanding of science in the nineteenth century.

Harriet Knight completed a PhD dissertation entitled 'Organising Natural Knowledge in the Seventeenth Century: the Works of Robert Boyle' at Birkbeck College, London, in 2003. She is currently an Arts and Humanities Research Board postdoctoral research fellow on the Oxford Francis Bacon project, based at the AHRB's Centre for Editing Lives and Letters, Queen Mary, University of London.

Susan McMahon received her PhD in 2001 from the University of Alberta, Edmonton, Canada, and recently held a postdoctoral fellowship at the Max Planck Institute for the History of Science in Berlin. She is currently working on a book about early modern natural history in England.

Richard Somerset is a *maître de conférences* at the Université de Nancy 2. He is particularly interested in the bridges between scientific and non-scientific treatments of temporality, especially in the nineteenth century. His publications have dealt with the conceptual underpinnings of the writings of such figures as Victor Hugo, Jules Michelet, Honoré de Balzac, George Eliot and Ralph Waldo Emerson.

Momme von Sydow is currently researching Bayesian models of reasoning and is attached to Georg-August-University's psychological DFG 'categorisation and induction' project in Göttingen, Germany. He has studied Philosophy and Psychology at the Universität Bonn (including courses in Biology) and his doctoral thesis, 'Sociobiology, Universal Darwinism and Their Transcendence', was written under the supervision of Professor David M. Knight in the University of Durham's Department of Philosophy.

Series Editor's Preface

The essays in this volume take us through two centuries of the varied interactions of science and beliefs, especially religious beliefs in England – although other parts of the British Isles make their appearance. The interactions have at times been wholly fruitful, and at other times more destructive; but that science and beliefs have repeatedly modified one another emerges clearly here.

Robert Boyle was very much a Christian virtuoso, for whom natural philosophy (including chemistry) was a virtuous and religious activity. Isaac Newton argued that to treat of God, arguing from the evidence of phenomena, was rightly a part of natural philosophy. The first of the Boyle Lectures presented Newton's mutual reinforcement of science and religion, while Gottfried Wilhelm Leibniz in Hanover asserted that natural religion was decaying in England, and that Isaac Newton's natural philosophy was the culprit. Belief in science and belief in God influenced one another, and both sets of belief were influenced by local cultures and institutions. Almost two centuries after Newton, however, Charles Darwin bleakly contemplated the extinction of all life on earth when the sun burned out, and wrestled with his own agnosticism and atheism. Thomas Henry Huxley trumpeted the downfall of theologians. On both sides of the Atlantic, the rhetoric of warfare was applied to the perceived clash between the natural sciences, successors to natural philosophy, and religion. Galileo's expressed hope that the two true books, nature and the bible, would ultimately agree with one another, seemed to many to have decayed between the seventeenth and the late nineteenth centuries. We see in this volume that some blamed the spread of cheap secular books, an influence soon countered by the publication of cheap and useful Christian volumes. Cardinal Newman held that unbelief was unavoidable in an age of intellect. Theologians as well as scientists explored the nature of evidence. It was their exploration that, in the last decades of the nineteenth century, helped, as Owen Chadwick has shown, to dissipate the illusion 'that science or philosophy were not only changing religious attitudes but strangling religious beliefs'.[1] We see here that natural theology outlasted Darwin and his century.

Neither science nor religion has ever been monolithic. One of the insights developed explicitly and implicitly throughout this book is that particular sciences, including natural history, zoology, geology, physics, and chemistry, have contributed to different approaches to religious belief. Another insight, raised unambiguously in the title of this book, *Science and Beliefs*, is that there are many different kinds of belief that interact with science – moral, philosophical, theological, and more besides. Some believed with

---

[1] Owen Chadwick, *The Secularization of the European Mind in the Nineteenth Century*, Cambridge, 1975, 228.

John Stuart Mill that natural science could provide a foundation for ethical behaviour. But it is important to recognise, as we see here, that the various sciences have their own foundations of belief, and that there is a continuing dialogue, or rather there are many continuing dialogues, between scientific and religious beliefs – each with their own canons of evidence. Contexts are important too, cultural, societal, and professional. Huxley was among the leaders of the struggle for professional status for scientists – 'jobs for the boys'. Professionalism of a different kind had long operated in the Royal Navy, where scientific observations and collections were carried out in the context of naval discipline. And whereas Huxley was competing with clergymen in established universities or country parishes, there was no competition at all between science and religion in the Navy. Indeed, as we see in these essays, the relations between science and religious and other beliefs in the space of 200 years have been both varied and very often mutually supportive. The clash between Huxley and Wilberforce at the British Association's meeting in Oxford has come to have iconic and almost mythological stature; but contemporary perceptions were far different. One of the many strengths of this set of inter-related essays is the way in which it encourages more nuanced perception and revisionist historiography.

Trevor Levere
University of Toronto

INTRODUCTION

Chapter 1

Science and Beliefs

David M. Knight

Introduction

In the summer of 1960 I was at Oxford, just finishing my degree examinations in chemistry and wondering how to spend my fourth year – Oxford being then a pioneer in the system of requiring from science undergraduates a year's work on a research project before graduation. My apprenticeship to science, involving boring corks, blowing glass, and watching, sucking, sniffing, touching and even occasionally tasting odd things,[1] had been character-building and sometimes enthralling: but I knew that for me it was time to stop, as it had been for Oliver Sacks much earlier in his life.[2] My sympathy and capacity for mathematics and logic were not such as to make a purely theoretical project possible. So I was sent along to have a word with Alistair Crombie, who taught History of Science, and had recently arranged for refugees from the laboratory to spend their year under his supervision. At that point, I had little feeling for history of chemistry: we had had some lectures on the discovery of the 'rare earth' (lanthanide) metals, but this confirmed the notion of most of us that such antiquarian activity should be left to the superannuated. But Alistair was sympathetic and listened: and when I said that I supposed I'd like to know what had made some great scientist tick, suggested that I should work on Humphry Davy's electro-chemical researches.

Alistair was an expert on the later Middle Ages and the Renaissance, and by training (in Australia) a physiologist: so his supervision was to be by benign neglect, an excellent system when it works. He put me straight about primary and secondary sources, told me to keep notes on sheets of paper of the same size, and advised me when writing not to quote too much. He gave me an introduction to the distinguished retired engineer Sir Harold Hartley, who had grown up in the nineteenth century and was wonderfully helpful to me as to other young historians of science; and who was himself working on Davy, writing a biography,[3] which was an excellent book when

---

[1]See my 'Then … and Now', in P. Morris, ed., *From Classical to Modern Chemistry: the Instrumental Revolution*, London, 2002, and 'Orientations', *Philosophical Writings*, 14 (2000), 67–72.

[2] O. Sacks, *Uncle Tungsten: Memories of a Chemical Boyhood*, London, 2001.

[3] H. Hartley, *Humphry Davy*, London, 1966.

it came out. But Sir Harold was concerned lest I made science out to be an intellectual game (Alistair was sure that its history was a part of general intellectual history, after all); and confessed to me that he could make nothing of Davy's posthumously published *Consolations in Travel.* To have a mentor who gave warnings like that, and who drew attention to what he wasn't interested in, was a great and rare boon. Enthralled, I found myself one day in Green Park, walking like Prospero a turn or two to still my beating mind, after finding something in the Royal Institution's library that confirmed a hunch about Davy's beliefs concerning matter. I duly wrote an undergraduate dissertation, and was fortunate enough to be awarded a studentship for three years, to take a Diploma in History and Philosophy of Science and thus complete my basic education, and to write a doctoral thesis on chemical atomic theory in the first 70 years or so of the nineteenth century. I was hooked; and at the end of that time I was extraordinarily lucky again in getting a new lectureship at Durham.

What's History of Science About, Then?

In 1960, the Scientific Revolution of the seventeenth century, about which Herbert Butterfield had rung the bell that called such wits as Alexandre Koyré, Bernard Cohen, Rupert Hall, Derek Whiteside and others together, was the central preoccupation of historians of science, then in close alliance very often with philosophers: history and philosophy of science at that time generally seemed one subject, with different aspects, rather than Siamese twins crying out to be separated (the enthusiastic Rom Harré in Oxford exemplified this). Alistair had indeed been engaged in pushing the start of the revolution back, all the way to the twelfth-century renaissance; and behind that was the science of the ancient Greeks, and the Babylonians whose theory-free mathematical devices were particularly appealing when logical positivism reigned, especially in Oxford, which was then the centre of things in Anglo-Saxon philosophy. Astronomy was thus especially attractive: it seemed to have made the running, provided the model of what a science should be like, and it had certainly aroused the ambitions of bright young men like Davy, for example, to be the Newton of chemistry.

In fact, the history of chemistry like the history of medicine was still largely in the hands of retired practitioners, who had not been warned of the dangers of Whig history (that invention of the Scottish Enlightenment[4]), and whom we young Turks despised, with the confidence of youth and professionalism, as amateurs. Allen Debus was a lone voice pointing to the enormous importance of Paracelsus, not nearly so

---

[4] M.G.H. Pittock, 'Historiography', in A. Brodie, ed., *The Cambridge Companion to the Scottish Enlightenment*, Cambridge, 2003, 258–79.

welcome in the pantheon as William Harvey or Robert Boyle (who was perceived, in the eighteenth-century manner, as a mechanical philosopher), and to medicine as the nursemaid of chemistry as well as other sciences. Important things, as we began to see, had no doubt got lost in the received view. But another feature of concentrating on the Scientific Revolution was that it seemed to indicate that the great men of that time, like the Fathers of the Church or the Founding Fathers of the USA, had laid down a pattern that would with minor amendment endure for ever. A scientific understanding of the world would follow from the proper application of method worked out by Isaac Newton and his predecessors: they left queries, but had sketched the picture which subsequent generations, petty men peeping about in their shadow, would fill in and colour. The nineteenth century would therefore be a bit dull, a bit of a come down; and yet those of us busy there knew it wasn't.

Then, in July 1961 under Alistair's auspices a great conference was held in Oxford under the title 'Scientific Change'. There, Thomas Kuhn, whose *Structure of Scientific Revolutions* was written but not yet published, gave a talk which electrified us graduate students, there as dogbodies to carry briefcases, check on the magic lanterns, and shepherd the eminent professors into the dining hall or lecture theatre. He spoke about dogma, and its essential role in science: his 'normal science' was indeed closer than many wished to painting by numbers; but that was not the most important thing. Kuhn had a new way of looking at belief, bringing the church scientific into line with other churches. Science was not some disinterested and isolated search after truth, free of metaphysics and manifestly a good thing in a naughty world: it was one human activity among others, not an austere model of rationality. Moreover, there had not been just one Scientific Revolution: the history of science was like the history of France, prone to revolutions (there had been one there recently, after all, leading to the Fifth Republic). Each Scientific Revolution was associated with new beliefs about the world, and with new language – for what might have seemed positive knowledge was theory laden, and only made its way through rhetoric or other (maybe more sinister) means of persuasion as the scientific community (or its younger members) was converted to the new way of seeing and believing. Science and beliefs were inseparable. This was very promising for us of the new generation, moving forward in time from the epoch of Newton and finding a new world of specialists.

Moving Forward into the Nineteenth Century

A lot had been written about the science of the nineteenth century, especially with the centenary of Charles Darwin's *Origin of Species* in 1959; but much of this had been done from a twentieth-century perspective, looking back and finding precursors or prematurity[5] rather than coming forward and therefore focusing on scientific change, and thus on a fuller context. Those of us coming from history and philosophy of science, much stimulated in Oxford by Rom, found with delight John Theodore Merz's *History of European Thought in the Nineteenth Century.* The first two volumes of this mighty work, freighted with footnotes which often dwarf the text, were concerned with science and had been written just as the nineteenth century was ending by a well-educated electrical engineer (whose splendid library is now in the University of Newcastle). Cutting across ordinary categories, mixing institutional history with intellectual and personal, this was an escape from history after the fashion of the scientific review-article, with its bloodless manner, passive voice, and mainstream, progressive focus. Merz, in his way, did history of science in the grand manner. Others must have rediscovered him also, because in 1965 he was reprinted in paperback. There are ways in which later writing like mine on nineteenth-century science is (further) footnotes to Merz.

I read Merz's work in the library of the Oxford Union, in faded Victorian splendour where one could peruse crumbling books while seated in crumbling leather-covered chairs with brass bookrests, toasted by a brazen electric fire. And there also were William Whewell's volumes on the *History* and the *Philosophy of the Inductive Sciences*, with categories again cutting across the taxonomies that I had got used to, and powerfully underlining the connections of science with big ideas, and with beliefs about the world. There, too, I discovered the *Bridgewater Treatises*; but at that point I had also been reading some Karl Barth,[6] and while I inwardly digested William Prout's volume because it contained his notions about atoms (along with the function of digestion, and meteorology), the whole project of natural theology seemed Panglossian. It was hard at that time to enter into that context, although its importance was more and more clearly underlined[7] as I read more science, especially that written for, or spoken to, a general audience. William Kirby's curious Trinitarian natural history (like the Quinarian system of William Swainson, his fellow disciple of William Jones of Nayland) made his *Bridgewater Treatise* particularly odd; Charles Babbage's unauthorised volume was odd in a different way; William Buckland's fold-out frontispiece, over a metre long to indicate the extent of geological history, was striking; but mostly they were not, I felt, for me. Most people felt the same, and when belatedly I did feel like acquiring copies they could be picked up quite cheaply.

---

[5] E.B. Hook, ed., *Prematurity in Scientific Discovery: on Resistance and Neglect*, Berkeley CA, 2002.

[6] Cf. P. Bowler, *Reconciling Science and Religion: the Debate in Early Twentieth-century Britain*, Chicago, 2001, 283–4.

[7] Cf. J. Brooke and G. Cantor, *Reconstructing Nature: the Engagement of Science and Religion*, Edinburgh, 1998, 141–75.

That interest in what made people tick led towards the region of what was then seen as mere biography, lending at best human interest and verisimilitude to what would otherwise have been a bald tale of experiment and inference as the truth gained ground upon the darkness. As an undergraduate, I got the chance to hold forth to engineers at the Polytechnic, now Oxford Brookes University. When (on leaving the Royal Artillery) I went up to Oxford in 1957, Sputnik had just been launched, and Charles Snow had delivered his famous lecture on 'Two Cultures': these things meant that there was lots of money for higher education, especially in science, and great concern that scientists and engineers should have a liberal element in their training. This one hour a week was what I was supposed to deliver: I found I enjoyed being on my hind legs, and discovered that the students (reluctant heroes, sleepy at the end of a long day) responded best to history with the people left in, and their hopes and beliefs as far as possible explored. Similarly, Durham University, stimulated by Stephen Toulmin down the road at Leeds, had decided to have a historian of science to go with their philosopher of science in the Philosophy department, but had very little idea of what such a person would actually do: it was wonderful to have such a free hand.

Teaching History of Science

For many years, all the history of science there was formed a part of courses in History and Philosophy of Science, shared with Barry Gower; and we worked out systems of lectures and tutorials, with students presenting essays, where the generality of philosophy and the particularity of history were (we hoped) kept in balance. We attracted a number of students from the science faculty, and when the university in the 1980s launched a Natural Sciences degree we got more students, and often very good ones indeed. For that degree we needed a further, new course – or module, as they came to be called as the ideas of le Corbusier entered the world of educational theory, that resting place of obsolete notions. I had taken over from a colleague a course of lectures delivered to students in classics and theology, who thus got history of science in place of history of philosophy: but as departmental boundaries hardened with new budgeting systems, such interdisciplinary activity became more difficult – other departments couldn't be allowed such freeloading. But those lectures, on the boundary of history of science and history of ideas, looked like a good basis for a course on science and religion, where the nineteenth century would be the focus. It would be a long nineteenth century, because the French Revolution of 1789 seemed to mark the moment when, for large numbers of people, organised religion seemed to be an enemy of human flourishing, rather than something which, though undoubtedly needing occasional reform, was purely good.

Twenty years ago (and still I fear now) popular conceptions of science and religion in the nineteenth century focused (despite Owen Chadwick[8]) upon Charles Darwin, and his supposed huge and successful assault upon religious belief. My hope in starting my course was to put Darwin in his place. Students still gasp delightedly at the suggestion that the bicycle was probably more important than Darwin in undermining churchgoing, because it gave people something to do instead that was more fun. They come through reading material on public health to realise that trained and organised common sense could be as important as high-faluting theory: whitewash, sewers, filters and watermains worked in dealing with cholera, where days of penitence and fasting did not. The belief became general that evils can be dealt with by practical activity under the guidance of common-sense science rather than endured with Christian fortitude. It was easy to appreciate that bad smells and dirty water could not be good for you, and, indeed, Buckland on being appointed dean of Westminster put much energy into reforming the abbey's drains: but it diminished the role of Providence. A course on science and religion could begin with Joseph Priestley's Christian materialism, look at Davy's dualistic chemistry and those curious *Consolations in Travel*, investigate phrenology, go indeed into palaeontology and evolution, look at energy and entropy, and finish up with psychical research, and the uncertain new physics of the early twentieth century, when science was perceived to rest like other human activities upon faith.

Thomas Henry Huxley had referred to the church scientific, and the New Reformation: and those of us who were in Britain becoming the first generation of professional historians of science, taking ourselves and our journals seriously,[9] could not but be intrigued by the manner in which men of science had earlier come to see themselves as professionals,[10] experts, demanding support and career structures, and claiming authority, as the clergy, lawyers and doctors had done long before. Here our guide was Jack Morrell; and his great work with Arnold Thackray, *Gentlemen of Science* (1981), written to mark the sesquicentenary of the British Association for the Advancement of Science (BAAS), was a great beacon. Institutions, though usually of little philosophical interest, were clearly crucial in the development of science: and they embodied beliefs and hopes, quite different in the Parisian Academy of Sciences,[11] the Royal Society, the Royal Institution,[12] and BAAS, for example. Martin Rudwick's

---

[8] O. Chadwick, *The Victorian Church*, vol. 2, London, 1970, 1–23.

[9] M. Beretta, C. Pogliano and P. Redondi, eds, *Journals and the History of Science*, Florence, 1999.

[10] C.A. Russell, N.G. Coley and G.K. Roberts, *Chemists by Profession: the Origins and Rise of the Royal Institute of Chemistry*, Milton Keynes, 1977.

[11] M.P. Crosland, *Science under Control: the French Academy of Sciences, 1795–1914*, Cambridge, 1992.

[12] F.A.J.L. James, ed., *The Common Purposes of Life: Science and Society at the Royal Institution of Great Britain*, Aldershot, 2002.

very rich study of the early years of the Geological Society of London[13] demonstrated the beliefs and values at the centre of things. Science was public knowledge, and carrying it on was a social activity. It also had strong political connections. The career of Sir Joseph Banks, not distinguished for any great scientific discovery, assumed its due importance again as we appreciated that science administrators are crucial, and that committees, networking and correspondence are vital elements of science.[14] That other great correspondent, Charles Darwin, who escaped through marriage and ill-health from administration into research, has had his life and times magnificently illuminated by the publication of his letters[15] in a sumptuous edition.

Students taking this course, coming mostly from science or from philosophy, had to wrestle with context; and they mostly enjoyed it. They had to try to get inside the beliefs of people from a past that was very different from the present, and yet had led inexorably to it. And concentration upon religion and science proved also to be epidemic, in that not many years after my module had been running the Templeton Foundation began giving grants and awards for such courses. For me, this opened the way to workshops in Chicago and Toronto, where I learned a lot but was confirmed in my historical approach: the twentieth century, with its rise of the new phenomenon of fundamentalism and the establishment of neo-Darwinian orthodoxy, can be best understood by coming at it from behind. We are duly grateful to the Templeton Foundation and to the Wellcome Trust for funding for the conference on which this book is based. Having taught about science and religion in the nineteenth century, and learned much from students as well as workshops, the time seemed ripe for a book about it; and my *Science and Spirituality: the Volatile Connection* is just published.[16]

The success of the Natural Sciences degree led us to launch another module taught by me, on the alleged rise of two cultures between 1750 and 1870. This involves students looking at Linnaeus' descriptive ideal of science, just as important in what we call the Enlightenment as Newton's mathematical and experimental way. We have to look hard at language, at Lavoisier's chemical revolution, and also at the didactic poetry of Erasmus Darwin with its footnotes and endnotes, mocked by his political opponents in the *Anti-Jacobin* and contrasted to the *Lyrical Ballads* of William Wordsworth and Samuel Taylor Coleridge,[17] and the writings of their romantic successors. This leads

---

13 M. Rudwick, *The Great Devonian Controversy: the Shaping of Scientific Knowledge among Gentlemanly Specialists*, Chicago, 1985.

14 N. Chambers, ed., *The Letters of Sir Joseph Banks: a Selection, 1768–1820*, London, 2000.

15 F. Burckhardt, et al., eds, *The Correspondence of Charles Darwin*, Cambridge, 1985 onwards.

16 D.M. Knight, *Science and Spirituality: the Volatile Connection*, London, 2003.

17 T.H. Levere, *Poetry Realized in Nature: Samuel Taylor Coleridge and Early Nineteenth-century Science*, Cambridge, 1981.

us to novels, notably Goethe's *Elective Affinities* where the characters are learning the chemistry of the double-decomposition reactions echoed in their own lives: making the reader uncertain whether to believe that people have no more choice than chemical elements, or are completely different. And Mary Shelley's *Frankenstein* is similarly open to different readings, but is clearly articulating beliefs about science very different from those of Davy, whose Royal Institution lectures are paralleled by those of Professor Waldman which first catch young Victor Frankenstein's attention in the book in which he ends up as victim. We look also at other more straightforward pieces of popular science, conveying different messages – and by no means always those to which the eminent professors of the day would have assented.

Institutions feature here also, notably exhibitions like that in the Crystal Palace, and museums; but also schools, colleges and universities. The end date of 1870 marked the Franco-Prussian war, where the better-educated nation, to the surprise of most commentators, defeated the apparently more warlike Second Empire of Napoleon III. At this time, compulsory primary education came belatedly to England; Oxford and Cambridge were in the throes of modernising their curricula, abolishing religious tests and patchily adopting the research ethos of the German universities and provincial universities were at last under way. All this involved debate about liberal education and technical training, secularity, and the place of science (or maybe the study of scientific method) in education:[18] and in our highly specialised world, in which engineers or physicists may be fundamentalists, the way things went is clearly (as ever) not simply 'a good thing'.

Running right through all this has been for me the increasing conviction that science is a social and practical activity as well as an intellectual one. Beliefs are not abstract and untestable convictions, but concern how things should be done. When I began my career, there was a noisy controversy between internalists and externalists: it perhaps made some sense in the days when historians were close to philosophers (and history of philosophy still generally appeals to a narrow context) but now seems absurd; as perhaps does our unrequited wooing in the 1960s and 1970s of philosophers, sociologists, historians, and any other attractive creatures whom we might think of dating.[19] It didn't work: the marriages announced did not take place. Some would have been marriages of convenience, or were perhaps conceived pragmatically as dynastic alliances: but be that as it may, all these moves marked different perceptions and beliefs about the way history of science should be done, who our allies were, and what needed accounting for.

---

[18] W.H. Brock, *Science for All: Studies in the History of Victorian Science and Education*, Aldershot, 1996.

[19] See my 'History of Science in Britain: a Personal View' [1984], reprinted in my *Science in the Romantic Era*, Aldershot, 1998, 1–11.

Beliefs in and around Science

The conference was therefore devoted to science and beliefs generally, and this book begins with essays that deal with beliefs in and about science and its practitioners. We begin with Barry Gower, whose paper is general and philosophical, distinguishing in the manner of Wilhelm Ostwald a century ago the Romantic and the Classical temper in science. His is a more timeless way than mine, and while I am happy to concede that there have always been all sorts of reasons to study the natural world, and all sorts of scientific temperaments, I still find it better to think of the Romantic period rather than of Romantic people. This is clearly a continuation of friendly arguments from the many years when we collaborated happily in teaching History and Philosophy of Science, and illustrates our somewhat different beliefs about it, as we and our disciplines have changed – maybe evolved.

Barry's paper is followed by a series of essays which explore the transition from natural history and natural philosophy to what we call science. Harriet Knight and Susan McMahon deal in different ways with the reputations of Robert Boyle and John Ray, and how their work was perceived, tidied, and made canonical in the eighteenth century, when the expectations of men of science were different as natural philosophy and natural history[20] gave way to something more explanatory. Michael Honeybone carefully explores the newer view of science in the context of a provincial society, howbeit one whose members had some metropolitan links; and Matthew Eddy investigates the use of imagery from medicine, the success story of eighteenth-century Edinburgh university, in making sense of mineralogy, where descriptive natural history was not enough. Randolph Cock takes us into the nineteenth century, as scientific professions began to emerge, and the Royal Navy became an important patron, some of its officers something like professionals. Finally, here, Aileen Fyfe takes us into the world of publishing, exploring popular science in reviews and journals to see the reactions and expectations of outsiders. This is where natural theology (which has been in the background throughout) makes its appearance; and we are ready for a transition to religious and metaphysical concerns which to most in the twenty-first century would seem external to science.

We turn accordingly to papers concerned with scientific and other beliefs, metaphysical in the case of Richard Somerset exploring the question why Darwinism caught on so well in Britain, the USA and Germany – but not in France. He examines two historians to cast light on ideas of progress. Momme von Sydow follows Darwin's pilgrimage from orthodoxy to agnosticism. Geoffrey Cantor illuminates an entertaining episode in Michael Faraday's career, casting light on his claim that his science and

---

[20] N. Jardine, J.A. Secord and E.C. Spary, eds, *Cultures of Natural History*, Cambridge, 1996.

religion were distinct by a close investigation of context: the gushing approach of Byron's daughter, Ada Lovelace. Frank James presents a very carefully contextualised account of the Huxley–Wilberforce debate in Oxford in 1860, indicating that there are still things to say about that event, or non-event: in particular, he looks at the position and reputation of the BAAS, at the end of whose annual meeting the debate happened.

Thomas Dixon indicates how 'altruism' was invented by respectable unbelievers, as agnostics like Huxley proved that irreligion did not entail immorality. William Brock has found fascinating connections between the work of William Crookes, later president of the Royal Society, in physics and in psychical research, exploring the boundaries of respectable science in an era when the man of science as an explorer, and the role of imagination in science, were being emphasised in reaction to earlier Baconianism which could make him seem a kind of superior technician. Crookes, like Davy and Faraday, would have been uneasy about the term 'professional', and was happy to go in for popular lectures (which enhanced his reputation). Peter Bowler takes us into the twentieth century with the career of J. Arthur Thomson, whose professional reputation suffered because he was perceived by his peers to be neglecting research in favour of popularisation, and thus failed to command the esteem of specialists. But he felt a vocation to get across a rather old-fashioned anti-materialistic vision of science to outsiders, in the manner of many we will meet in this book. And finally John Brooke brings the work to a conclusion, pulling the threads together, as he did in the conference. The whole makes a timely and coherent collection, and a good read.

What Did Make Davy Tick, Then?

This has turned out, as no doubt Alistair foresaw, to be the sort of question that has no final answer. Can one say, after 40 years of reading:[21]

> So word by word, and line by line,
> The dead man touch'd me from the past
> And all at once it seemed at last
> The living soul was flash'd on mine.

Clearly, as his first biographer John Ayrton Paris perceived and as Adrian Desmond saw for Huxley,[22] social mobility was the best key to his life. If not exactly from rags to riches, or log cabin to White House, his life showed how a man from very humble and provincial origins could through science end up hobnobbing with the

---

[21] A.Tennyson, *In Memoriam* (ed. S. Shatto and M. Shaw), Oxford, 1982, 112.
[22] A. Desmond, *Huxley: the Devil's Disciple*, London, 1995.

mighty, having his portrait painted by Thomas Lawrence, and being commemorated in Westminster Abbey. Like the medieval church, the church scientific could lead to an astonishing career. But, like the church, it entailed beliefs: about science, about politics, about spirituality. Like Banks, and like most of his contemporaries, Davy held Baconian views about science, emphasising utility and careful inductive reasoning from experiment. But, despite this proclaimed allegiance, his greatest experiment was not the 'suck it and see' he did with laughing gas, but his proof against the apparent evidence that water is decomposed by an electric current into oxygen and hydrogen only. He persisted, using apparatus of silver, gold and agate to exclude impurities, until nature, thus on the rack, gave him the answer he required. He laid hold on the idea that electricity and chemical affinity were manifestations of one power, and set about proving this belief; in the manner more of Whewell than of Bacon. In politics, the radical views he held in Bristol when associating with left-wingers like Thomas Beddoes and S.T. Coleridge were soon abandoned in London, where, lecturing to the wealthy about the potential of applied science, he declared his faith that the unequal division of property was the foundation of scientific and technical progress. And in religion, the orthodox Anglican young man became in Bristol a sceptic and materialist, later to renounce these doctrines in favour of a pantheistic dualism, coupled with an unusual sympathy in a man of science at that time for the Roman Catholic Church, which he knew in Ireland, Italy and Slovenia. These beliefs and his scientific activity are intimately but unstraightforwardly connected.

Biography, like science and probably like religion, involves a good deal of projection[23] – which is why my vision[24] differed from Hartley's:

> O Lady, we receive but what we give,
> And in our life alone does Nature live:
> Ours is her wedding garment, ours her shroud!

Nevertheless, it has like science at least a strong connection with reality and remains worthwhile, because it helps us to see that science is not some great entity that might interact with other abstractions like religion, but is a human activity carried on by people living at a specific place and time, working together to understand and improve the world. And what the essays in this book show is that conceptions and expectations of science have been very various, and that it never has been and cannot be some value-free activity, the positive knowledge for which Auguste Comte and his disciples hoped. It has to be based, as Arthur Balfour pointed out long ago,[25] upon belief.

---

[23] S.T. Coleridge, *Selected Poems* (ed. R. Holmes), London, 1996, 180.

[24] D.M. Knight, *Humphry Davy: Science and Power*, 2nd edn, Cambridge, 1998.

[25] A.J. Balfour, *A Defence of Philosophic Doubt*, London, 1879, esp. 303.

PART I
BELIEFS WITHIN SCIENCE

Chapter 2

The Metaphysics of Science in the Romantic Era

Barry Gower

In a series of influential papers David Knight has argued that, in the late eighteenth century and the early nineteenth century, the physical sciences had a 'Romantic' aspect. Goethe's anti-Newtonian theory of colours, Oersted's late essays collected under the title *The Soul in Nature*, and Davy's *Consolations in Travel*, together with Coleridge's study of Naturphilosophie,[1] provide the material for his arguments, and help support his historiographic claim that, at least so far as the chemical revolution is concerned, social history will be more illuminating than philosophy. 'Romantic science,' he has said, is 'characteristic of a time rather than of a cast of mind.'[2] We understand such a science better by seeing it as a response to the social context in which it came to prominence, than by seeing it as the rational expression of a distinctive and wide-ranging conceptual revolution which created that social context. The 'cast of mind' that some might wish to use to delineate the effects of this conceptual revolution will inevitably supply a framework or procrustean bed into which the historical facts will have to fit, and perhaps be distorted to do so. If we want to know what really happened – what it really felt like to be a Romantic scientist – we should eschew the risks and prejudices involved in depending on frameworks, and other such creations of hindsight; we should look at what really happened and create our picture of Romantic science by setting it within its social environment. The argument is a familiar one: it concerns rational reconstructions versus historical narrative, and the issue of internal versus external historical analyses of science.

But the supposed distinction misleads us. All historical narrative involves reconstructing the past, and since the past involves human actors we should expect that reconstruction, and thus that narrative, to appeal to the norms of rationality. If Romantic science is characteristic of a time, it is so – at least in part – because the time in question found a particular way of articulating a cast of mind. Of course, those

---

[1] J.W. von Goethe, *Goethe's Theory of Colours*, trans. C.L. Eastlake, London, 1840; H.C. Oersted, *The Soul in Nature: With supplementary contributions*, trans. L. and J.R. Horner, London, 1852; H. Davy, *Consolations in Travel, or, the Last Days of a Philosopher*, London, 1830.

[2] D.M. Knight, *Science in the Romantic Era*, Aldershot, 1998, x.

historians who have examined the writings of Romantic scientists have often been very puzzled by what they have read. Davy's *Consolations*, for example, can seem like an attempt to express feelings rather than rational thought, like poetic fiction rather than scientific fact. And this appearance is reinforced when we learn that the Romantic scientists kept company with poets, and were sometimes themselves accomplished authors of poetic literature. Similarly, the German naturalist Lorenz Oken wrote in a way that has mystified his readers.[3] He used the distinctive idioms of biological language to express that revolution of sensibility, associated in Germany with the Schlegel brothers and Novalis, which celebrated the 'magic of the ineffable'.[4] Some of the bewilderment we feel when we read Davy or Oken is indeed dispelled when we understand what they wrote not as any kind of science, but as imaginative literature rooted in a specific social and cultural setting. But questions remain. For, Davy and Oken exemplify a more general trend that is by no means confined to the Romantic era. A good many scientists, some with the most distinguished reputations, seem to have been compelled to think and write in ways that their contemporaries and successors have found either difficult to comprehend or excessively speculative, or both. For what audiences are they writing? Of course, the answers vary from person to person and from time to time. But still the fact that there are similar questions suggests that a complete answer to any one of them will have to look beyond the particularities of time and place, and draw attention to the character of the intellectual need that scientists are trying, in their different ways, to meet by venturing into abstruse abstractions or elusive imagery.

To support my general view I will sketch an argument for the conclusion that some of the important features of a science we might call Romantic are of a type which occurs frequently in the history of science. I shall claim that these features have roots in an Aristotelian conception of metaphysics. This conception has often provided a powerful driver for the thinking of certain scientists, and it helps too to supply the context for those who see a need to integrate science into a single culture. The form of the metaphysics does indeed vary strikingly from one time to another, and to that extent it must be right to say that its particular characteristics are temporally determined, but there is nevertheless a distinctive cast of mind providing the framework for these intriguing but puzzling aspects of scientific thought.

Aristotle's conception requires that metaphysics, far from being idle speculation or instrumental presupposition, makes powerful claims about the world, including those aspects of the world science investigates. It differs from the view of those who take metaphysics to be just the unconscious presuppositions of a thinker or an age. That is a quick way of attributing metaphysics to a scientist, since scientists can be

---

[3] L. Oken, *Elements of Physiophilosophy*, trans. A. Tulk, London, 1847.

[4] M. Praz, *The Romantic Agony*, trans. A. Davidson, London, 1933, 33.

presumed to presuppose things, even if they do so unconsciously. And plenty of work has been done on making explicit the tacit assumptions particular scientists are supposed to have adopted. Talk of 'Weltanschauungen' and of intellectual environments and influences, is the product of such work, and although I do not wish to go so far as to deny that there are such environments and influences, I am sceptical about the extent of their coherence. In part my scepticism derives from the familiar difficulties that writers such as Kuhn have encountered in trying to articulate such claims clearly. Aristotle's conception is more discriminating; it accepts the presuppositional nature of metaphysics but claims that it is more explicit than is often recognised. It is presuppositional in the sense that it identifies and argues for 'first principles' of science. It constitutes, therefore, a distinct intellectual endeavour, with its own subject matter and a characteristic method. I do not assume that every scientist, even the most prominent, has a metaphysics in this sense. Indeed I assume that most do not, if only for the reason that they do not undertake the task of identifying and arguing for 'first principles'. The development of science often proceeds quite independently of metaphysics in this sense, and entirely to the satisfaction of those scientists most closely involved.[5]

What, though, is the case for a 'first philosophy'? The argument draws on some philosophical and logical ideas about scientific explanation. This will be helpful, for if there is a place where science and the spirit of Romanticism can meet it will be in the way scientific theory provides us with a deep and satisfying understanding of nature – a philosophy of nature if you will – that can and must inform a person's intellectual response to the world. Our concern, therefore, is with science as natural philosophy – as answering the need for, as Aristotle would have put it, an understanding of the world as consisting of physical, material beings. Such an understanding must be grounded in how the world really is, as opposed to how it appears from this or that point of view. But of course there is a question about whether the concept of the way the world really is makes sense. And this question – whether there is an ultimate reality – is the same as the question whether what we might call the explanatory regress is indefinite. Facts used to explain at one level are themselves explained by other facts at a more fundamental or deeper level, and these other facts might be explained by facts at a still more fundamental level, and so on. If this sequence of explanations is indefinite then scepticism about ultimate explanations and thus about ultimate reality would seem justified. We will have to do without the kind of understanding Aristotle hoped that natural philosophy would provide. We will have knowledge of the facts, but not reasoned knowledge of them.

---

[5] For this conception of metaphysics, see G. Hatfield, 'Metaphysics and the New Science', in D.C. Lindberg and R.S. Westman, eds, *Reappraisals of the Scientific Revolution*, Cambridge, 1990, 94–5.

In favour of the view that the explanatory sequence cannot be indefinite we might urge that if it were indefinite no facts would explain any other facts because no explanation would be properly complete. This is what is meant by the claim that an infinite explanatory regress would be vicious; explanation is impossible if there is an infinite explanatory regress. But although this completeness requirement is widely accepted and was explicitly adopted by such philosophers as Plato, Aristotle and Mill,[6] philosophers of science nowadays regularly dismiss it as mistaken: something can be a complete and full explanation of something else, irrespective of whether there is an explanation of that something, and certainly irrespective of whether we can explain it. This dismissal is, I think, entirely legitimate if we think of explanation in causal terms. If something is, or contains, the cause of something else, and for that reason explains that something else, then there is no need to pursue a causal regress in order to ensure that it is a full and complete causal explanation of that something else. And we do not need to know what causes something in order to use it as (part of) the causal explanation of something else. But if we think of explanation as subsumption under law then the rejection of the completeness requirement is questionable. For to explain a fact by subsuming it is to show that it follows logically from other facts. We are required therefore to produce subsuming facts, and this means that the claims we use to report what we take to be those subsuming facts must be justified in some way. It is not implausible to suppose that the kind of justification required is explanation of these facts by subsuming them under other facts. In other words, a scientific law (e.g. Kepler's law that the planets move in elliptical orbits) cannot explain – by subsuming – a particular fact (e.g. why Mars is observed to travel in a particular manner in the night sky) unless there is a higher level scientific law which explains – by subsuming – that lower level law, and there is a still higher level law which explains – by subsuming – that high level law, etc. The degree to which we can explain a particular fact is then a matter of the extent to which we can justify the relevant lower level law by explaining it, and justify the relevant high level law by explaining it, and justify the still higher level law by explaining it, etc. If that means we must explain an infinite number of things in order to explain a particular fact to any degree than clearly we cannot explain a particular fact to any degree. But this conclusion is surely wrong: we can explain facts to some degree. And that means the explanation regress cannot go on for ever. Justifications must have foundations in ultimate explanations.

This is not yet a good argument; it needs, for instance, more careful analysis of the relations between explanations and justifications. That task must await another time, but let me here suggest that we can support the concept of ultimate explanations by considering the role of unification in explanation. For it has been argued that the point of explaining facts by subsuming them under more general facts is to explain by

[6] D.H. Ruben, *Explaining Explanation*, London, 1990, 125.

'achieving a unified vision of the phenomena'.[7] We need to be clear about what this means, and on one view – a realist view – explanation is a matter of giving an account of the constituents of nature and of their essential characteristics. From the principles embodying that account we derive laws of nature expressing natural necessities. The system of derivations that results reflects the structural organisation of nature itself. There will be as much unity in explanation as there is unity in nature. There is, in other words, only a contingent connection between explanation and unification. But if, as has been pointed out, nature is not unified then this realist view provides no justification for the unification requirement. On the other hand, if we adopt an anti-realist view, (metaphysical) explanation is eschewed and the quest for theoretical unification can only be justified as a heuristic maxim: prefer unified theories because they are easier to work with. Neither realism nor anti-realism gives us secure grounds on which to defend the view that explanation requires unification.

Kantian ideas do, however, enable us to provide that defence, and to propose an 'explanation argument' in support of the required conclusion. Our faculty of reason, Kant said, uses principles requiring 'systematic unity' in our knowledge. This unity is 'indispensable' because 'knowledge is to be not a mere contingent aggregate, but a system connected according to necessary laws'.[8] But the principles in question are regulative rather than constitutive. This means that systematic unity is 'only a *projected* unity' even though it serves – as it would if it were constitutive of nature – as 'the criterion of truth'.[9] Kant's point is that the quest for unity in scientific explanation has no objective force, in the sense that it is not driven by the belief that nature itself is unified – whatever that might mean. But this does not mean that the quest is driven by nothing other than the desire to make explanations conform to a subjective preference for unity. There is a *via media* between these familiar alternatives. The quest for unifying explanations does have some objective validity, though we need to understand this to mean that only with the aid of a principle of unity, projected though it is, is objective knowledge possible. The unity we seek in explanation is not 'out there' in nature, and we are simply representing this feature of reality in explanations. Nor is the unity to be understood as imposed on an otherwise disunified nature by the action of a God whose existence would first have to be postulated. Rather, unifying explanations are the outcome of adopting a methodological or regulative principle requiring unity if the knowledge expressed in scientific explanations is to be possible. It is regulative rather than constitutive because reason does not assume or postulate the existence of unity in nature, but 'requires us to seek for … unity'.[10] The unity

---

[7] P. Kitcher, *The Advancement of Science: Science without legend, objectivity without illusions*, Oxford, 1993, 171.

[8] I. Kant, *Critique of Pure Reason*, trans. N. Kemp Smith, London, 1964, A644–A645.

[9] *Ibid.*, A647.

[10] *Ibid.*, A651.

is, then, 'transcendental' rather than 'empirical'. It is not merely a heuristic aid but a means of explaining by recording objective relations, necessities and dependencies in nature. Of course there is a sense in which a projected unity may not correctly represent what nature is like 'in itself', but Kant's critical philosophy was intended to show that such a possibility is idle. The principle which says that nature is unified is not a substantive claim about nature which a realist might use to justify the methodological preference for theoretical unification, for that simply presents us with the insuperable epistemological problem of justifying the claim, and thus the methodological preference. It is, rather, a regulative principle enjoining us to practice science in a particular way, namely by seeking theoretical unification. What, though, is the force of this principle? Why should we follow it? In brief, Kant's claim is that we must follow it in so far as we are concerned, in explanation, in securing truth. Our goal in explanation is truth – correct explanations; to secure that goal we must seek explanations which integrate phenomena into a unified system.

There are some important and controversial details about this line of thought. I would not wish to place too much weight on the support that such an account of the relation between unification and explanation can offer to the view that there are ultimate explanations and that such explanations provide us with an absolute conception of reality. I think though that it does provide some support. In particular it suggests that scepticism about unification, ultimate explanation and an absolute conception of reality is indeed justified within the framework of scientific realism, for there we will always have to recognise a distinction between what is acceptable and what is correct. Kant's transcendental idealism closes the gap, as indeed do the views of more recent writers such as Peirce and Putnam.

What form would ultimate explanations take, and what would make them ultimate? It is at this point that I turn to Aristotle's conception of metaphysics. For him, its subject matter was 'being *qua* being', and he gave the name 'first philosophy' to the science concerned with it. Sometimes, and significantly, the Greek word for 'theology' was used as an alternative to 'first philosophy'. There are, he said, two other sciences in theoretical philosophy, namely mathematics and physics (or natural philosophy). All three of these 'sciences' study being in one way or another; they have the same subject matter, being – or existence if we use that term generously and apply it not just to physical existence. The distinction between the three parts of theoretical philosophy rests upon differences between the ways in which they study being. First philosophy is concerned with those characteristics or aspects of being, if any, which would be characteristics of being *qua* being; what has to be true of numbers, or properties and relations, or minds, or physical objects, if they are to be, or to exist? Mathematics is concerned with those characteristics of being, if any, which would be characteristics of being *qua* mathematical object, or of existing objects in so far as they are mathematical. And physics is concerned with those characteristics of being, if any, which would be characteristics of being *qua* physical object, or of existing objects in so far as they are physical. So, like mathematics and physics, metaphysics does not need there to be special sorts of things for it to study, but it does need there to be special sorts of properties of ordinary sorts of things. The three parts of theoretical philosophy, then, complement rather than compete with each other. They are concerned with different specific sorts of properties of objects, however widely or narrowly we take the domain of objects to be: mathematics is concerned with the mathematical properties of objects, such as the property of being a triangle, and natural philosophy with the physical properties of objects, such as the property of being an elephant. We might express an idea similar to that of Aristotle by saying that a part of mathematics should be concerned with the kind of being, or way of being, that is peculiar to the objects of mathematical study (e.g. numbers); and a part of natural philosophy should be concerned with the kind of being, or way of being, that is peculiar to the objects of physical enquiry (material, physical). The idea is that the way of being that is peculiar to mathematical objects, such as triangles or prime numbers, is quite different to the way of being that is peculiar to physical objects, such as electrons or elephants. Thus far we might agree with Aristotle, or at least see what sort of position he is developing, even if we do not want to occupy it ourselves. But the next step takes us into mysterious territory. For the line of thought being developed leads to the conclusion that metaphysics will be concerned with the kind of being, or way of being, that is peculiar to being *qua* being. We can readily produce a negative characterisation of metaphysics' concern: it is concerned with real things, but not with them in so far as they are physical or mathematical, for there is a way of being of real things which is not that which is characteristic of either physical things or mathematical things. A positive characterisation begins to emerge when we recall that Aristotle understood metaphysics as *first* philosophy. This implies that its concerns are those taken for granted by the special sciences, mathematics and natural philosophy. In a general way we can express this idea by saying that in so far as the special sciences have a demonstrative structure they will depend on axioms or theoretical principles which, because they are not themselves demonstrated, or proved or explained, are taken for granted or assumed. More specifically, metaphysics must study those properties or aspects of being which are taken for granted by the special sciences. It is concerned with how the subject matter of, say, a physical enquiry must be if it is to be real and objective. For example, physics claims that an essential property of gold is its atomic weight, and on the basis of that claim it explains why, for example, gold is yellow at normal temperatures and pressures. And this claim is intended as a report about how reality is, and not just as a perspectival belief about how gold appears to human beings. For another example, geometry claims that parallel straight lines do not meet, and on the basis of this claim together with other assumptions, geometers prove theorems about circles, right-angled triangles, etc. The geometers' claim is intended as a report about how space really is, and not just as a truth about how space appears to us. What, asks the metaphysician, must be true of the subject matters that physicists and geometers study if the objectivity with which they invest their claims is to be possible? What has to be true of gold, or of space, if they are really and objectively to have the essential properties that physicists and geometers claim them to have? In general, how must things be if we are to have objective scientific knowledge; knowledge, as Aristotle expressed it, of the 'reasoned fact'. To answer these questions, we must undertake a study of being *qua* being.

The outcome of such a study, if it is possible, will be claims which provide an absolute conception of ultimate reality. The being *qua* being of Aristotle's first philosophy is the ultimate reality of our metaphysics. Like the physical sciences, metaphysics studies reality, but studies different aspects of it. It is concerned with a way of being real that is not the same as the way of being physically or materially real which is the proper concern of the sciences. But metaphysics is logically prior to the sciences. Somehow the investigations of the scientists are founded on, or grounded in, the investigations of the metaphysicians. Ultimate reality is not the same as physical or material reality.

But what sort of study could this be? If it is to be a science like mathematics and natural philosophy then it will share their demonstrative structure. It will depend, therefore, on its own axioms or theoretical principles and will be unable to demonstrate their truth. But since it lacks that ability it will not be *first* philosophy. Somehow metaphysics must find a way of grounding its own assumptions as true and objective, without making use of the methods we customarily use to establish truth and objectivity. Aristotle's alternative to demonstration is dialectic, and dialectical method works by systematically examining common beliefs in order to achieve positive conclusions. For example, a dialectical examination of common beliefs about knowledge, or about explanation, results in a conclusion about what knowledge is, or what explanation is. Can we use such a method to secure the principles of first philosophy and thus exempt this science, uniquely, from the need to presuppose another science whose task would be the establishment of those principles? Can we, in short, give an account of the arguments required in first philosophy – dialectical arguments – that satisfies the condition that they be scientific, even though they are not demonstrative. On the face of it the prospects are not good. Dialectical arguments, being arguments from common beliefs, are indeed concerned with what is true and what is false, but just because they rely on common beliefs they cannot yield knowledge. A dialectical examination of common beliefs can indeed be expected to result in a conclusion which is the best coherent representation of those beliefs. But that coherence is no guarantee that the conclusion is true. Objective knowledge, including the knowledge sought in first philosophy, must rest on principles which are better known than the conclusions drawn. But if the principles of first philosophy are derived using dialectical method, they will be simply matters of belief, not knowledge at all. There will be no genuine scientific knowledge in first philosophy.[11]

But suppose the dialectic concerns not just any and all common beliefs, but only those which we think are especially reliable. We will then have some reasons for believing in the objective truth of our conclusions; we will have 'strong' dialectic.[12] First philosophy, basing arguments on common beliefs selected for their reliability, will have some claim to be both scientific, in Aristotle's sense, and dialectic. What, though, should be our criteria of reliability? What makes a common belief of the relevant sort reliable? One suggestion is that it will be a presupposition of the special sciences – physics and mathematics. The 'strong' dialectic characterising first philosophy will be the consequence of an examination of such presuppositions. The reason why the presuppositions of the special sciences are not just common beliefs but form a particularly important kind of common belief is that they have an indispensability to them. If we were to abandon them then we would have to give up much else as well. We assume the reliability of these presuppositions. The appropriateness of presuppositions for strong dialectic is not, then, a matter of them being generally agreed. It stems from a more penetrating examination of what has to be true if we are to have knowledge of real things, in so far as those real things are physical, or in so far as they are mathematical. Using Aristotle's terminology, there has to be such a thing as being *qua* being if there is to be any scientific study of the objective world. The assumption or presupposition that there is such a thing as being *qua* being is, therefore, not just – or not even – a common belief. It has, rather, the status of a fundamental belief whose truth we take for granted, perhaps unconsciously, and which underpins a great many other beliefs.

Aristotle's dialectic is, of course, what we call philosophy. Potentially it suffers from the familiar and well-rehearsed problems of philosophical method. Philosophy seems to begin from common beliefs or what we would all say, and argues to conclusions about, say, the nature of reality. But the basis seems totally inadequate to authorise such conclusions. Why should common beliefs, or what we would say, count for so much? All we really have is a conclusion about the nature of reality which is consistent with the premises from which we have derived it, not a conclusion we would be justified in accepting as representing the way things are. And dialectic is no better off unless we can defend a form of it – strong dialectic – which will deflect this scepticism. Aristotle claims that first philosophy uses the distinctive method of strong dialectic to establish the objective presuppositions of the special sciences; we claim – or some of us do – that metaphysics uses distinctively philosophical methods to establish the truth and objectivity of what is taken for granted in the natural sciences. Once again Kant has provided us with the concepts and the arguments needed to ground

[11] T. Irwin, *Aristotle's First Principles*, Oxford, 1988, 146.
[12] *Ibid*., 176.

the Aristotelian project. In particular we put an epistemological spin on the quest for being *qua* being. Not: what do we have to assume there is for there to be objective scientific knowledge; rather: what are the a priori conditions of such knowledge. But nevertheless there is much congruity between Aristotle's conception of metaphysical science as being concerned with the establishment of those principles which are taken for granted by the special sciences, and Kant's claim that metaphysics 'treats of those elements and highest maxims which must form the basis of the very *possibility* of some sciences, and of the *use* of all'.[13] For both, metaphysics is a second-order enquiry. It examines the questions raised by the assumptions used by those who investigate nature, not the questions about the world that directly concern such investigators. For both, the philosophical/dialectical methods used in metaphysics have a legitimate claim to yield more than a 'likely story'.

The cast of mind I have tried to sketch has I think some potential in helping us to understand some puzzling aspects of the history of science, but also of scientific thinking more generally. In Newtonian studies, for example, historians have wanted to explain, or explain away, the embarrassment felt in the face of the enormous amount of time and energy Newton devoted to alchemy, to the origins of religion, and to 'ancient wisdom' more generally. What, they ask, was Newton trying to achieve in his extensive attempts to grasp and develop coherent conceptions of these matters? Of course, there is an element of illicit hindsight in such judgements, though the fact that Newton did not publish the results of his labours in these fields might be taken to suggest that he himself had misgivings. Part of the answer, I suggest, is that he was implicitly acknowledging the need for an Aristotelian first philosophy to underpin and indeed explain his mathematical principles of natural philosophy. The method appropriate to such a first philosophy is dialectical in the sense I have suggested. It uses, therefore, common beliefs. But if the dialectic is to be strong enough to sustain a credible first philosophy, the common beliefs to be used must be selected with care. What beliefs, of the relevant kind, would be appropriate? In answering this, we should not be at all surprised that Newton turned his attention to beliefs which he and his contemporaries were prepared to invest with authority and reliability. Religious beliefs satisfied that criterion and we have a good understanding of the role they played in his thought. So also did alchemical beliefs. In pre-modern Europe there was a strong Arabic tradition of alchemical thought which took alchemy to be the 'higher' of the two major divisions of 'worldly' knowledge.[14] And alchemy is generally said to have reached the height of its influence in the sixteenth century. Even more important for Newton was the recovery of ancient beliefs, particularly those associated with Pythagorean and Stoic thought. His writings show that he was firmly committed to

[13] Kant (1964), A851/B879.

[14] David C. Lindberg, ed., *Science in the Middle Ages*, 15.

the Renaissance view of history as degenerating from an original golden age when knowledge of the natural and supernatural world was uncontaminated by sin and error. So the appeal to beliefs held by ancient writers would satisfy the requirement that the dialectic of first philosophy should be strong. Work remains to be done, I think, on piecing together the complex elements of Newton's first philosophy – his Aristotelian metaphysics – though the recent work of some historians has given the outlines of the structure of ideas which links his studies in alchemy and in ancient cosmic thought with his designation of gravity as an 'active principle' and with the need for an explanation of the 'how' of gravity.

A more straightforward illustration of my cast of mind is found in those eighteenth-century Newtonians who tried to develop a 'dynamical doctrine of nature' to supplant the then prevalent mechanical atomism. Some, like John Michell, John Turberville Needham and Joseph Priestley were what we would call scientists, or natural philosophers. Others, like Robert Green, David Hartley and Francis Hutcheson were metaphysicians, or first philosophers. All, though, used an explicitly Newtonian dialectic, taking its cue from the 'Queries' appended to Newton's *Opticks*. They were in no doubt that his reputation was entirely adequate to the required authority of the common beliefs required for first philosophy. Needham, for example, sought to base a theory of biological generation on the idea of a 'vegetative force' produced by a combination of an active, expansive force and a force of resistance. For Green, 'action or force in general is the essence or substratum of matter'.[15] And in Hutcheson's moral psychology 'acts of the will may be ... divided into two classes, according as one is pursuing good for himself, and repelling the contrary, or pursuing good for others, and repelling evils which threaten them'.[16] This and other similar examples may look like nothing more than the opportunistic exploitation of a fashionable idiom, but I claim that the rhetoric serves a point of substance. There is a genuine attempt in these writers to identify a convincing and self-justifying conceptual bedrock for thinking in natural philosophy. From their point of view it was entirely natural and legitimate that they should use the authority of Newton to achieve this end.

But the rapid development of new areas of natural philosophy in the latter half of the eighteenth century made the appeal to Newtonian authority to legitimate the dialectic of a first philosophy using this dynamical doctrine of nature increasingly suspect. So let me turn, finally, to the rather different approach evident in some important aspects of science and philosophy in early nineteenth-century Germany – in so-called 'Naturphilosophie'. Here for the first time we see a new approach to

---

[15] R. Green, *The Principles of the Philosophy of the Expansive and Contractive Forces*, Cambridge, 1727, 286.

[16] F. Hutcheson, *A System of Moral Philosophy*, Glasgow and London, 1755, 8.

securing first philosophy – a move away from personal authority, whether ancient or modern, as a source of the strength of strong dialectic. It incorporated, of course, elements of the Romanticism prevalent in early nineteenth-century Germany, and it is those elements which give it its distinctive features. Naturphilosophie is Germany's Romantic science. But it also incorporated Kantian ideas and to that extent can count as an early attempt to use reason rather than authority as the source of the security and strength of its dialectic. As I have indicated, those Kantian ideas, however distorted and misunderstood they may have been, are at the heart of our own thinking about the best way to meet Aristotle's requirements for first philosophy.

In the metaphysics of Naturphilosophie, mind and world are inseparably linked; indeed they are in some sense one. For mind makes no sense without world and world makes no sense without mind. In particular, and this is where the idealism of the post-Kantians gets its grip, the concept of a mind-independent world is incoherent and must be rejected. This can be glossed in terms of the idea that the mind's understanding and knowledge of world is conditional on a projected conception of the world which makes understanding and knowledge of it possible. The features of this projected conception reflect the objectivity required for understanding and knowledge as well as the mind-dependent character of mind. In this sense the features are dialectical; they represent both 'thesis' and 'antithesis'. What the post-Kantians called the inner activity of nature is, therefore, polar in a generalised and abstract sense; its activity and productivity arises from the tension or conflict between opposites. Inert matter, for example, is to be understood as a product of this inner activity and we can conceptualise this as a physical analysis of matter in terms of attractive and repelling forces – as indeed Kant had done in presenting the dynamical doctrine of nature in his *Metaphysical Foundations of Natural Science*.[17] And, conveniently enough, this generalised polarity is just what is needed to ground an understanding of newly explored magnetic and electrical phenomena. In this post-Kantian first philosophy – this Naturphilosophie – we are not then considering a priori science; still less are we dealing with ambitious empirical theories. Rather, we should understand it as an attempt to meet the Aristotelian requirements for first philosophy by creating a dialectic – in both the broad and narrow senses of that term – with an authority derived from its use of beliefs that owed their security and strength not to ancient sages or illustrious moderns but to Kant's conviction that metaphysics is concerned with what is needed for the possibility of the special sciences, physics, chemistry, etc.

Let me conclude by relating my claim more explicitly to the point from which I began, namely David Knight's view that Romantic science is best understood as characteristic of its time rather than as expressing a cast of mind. I think we need both.

---

[17] I. Kant, *Metaphysical Foundations of Natural Science*, trans. J. Ellington, Indianapolis, 1970.

Of course we need an understanding of the social, and indeed intellectual, context in which the scientists we associate with the Romantic movement lived and worked. I do not wish to undervalue the illumination such an understanding provides. But cast of mind, as I have been interpreting that phrase, is also important. The Romantic scientists inherited a way of thinking about how science can provide understanding of our world; it had, and has, deep roots and is a key part of their, and our, cultural inheritance. We can enhance our picture of them and their doings by taking that way of thinking – that cast of mind – into account.

Chapter 3

Rearranging Seventeenth-Century Natural History into Natural Philosophy: Eighteenth-Century Editions of Boyle's Works

Harriet Knight

Introduction

The aim of this paper is to suggest a context for understanding the implications of the eighteenth-century production of epitomised, systematised editions of the works of Robert Boyle. I will suggest that Richard Boulton's 1699 epitome, and Peter Shaw's 1725 abridged, methodised edition of Boyle's works operate within the seventeenth-century conception of an inductive process whereby individuated 'historical' knowledge is transformed into systematic 'philosophical' knowledge through textual rearrangement. They thus represent material attempts to rescue Boyle's works from criticisms of it as insufficiently systematic to attain philosophical status, seeking to draw out the philosophical potential of his disorderly publications through textual manipulation. By contrast, Birch's chronological 1744 edition operates more in line with modern editorial practice, claiming to provide an accurate representation of the shape of Boyle's corpus, and not participating in the ambition actually to advance his work which is notably present in Shaw.

The Significance of Organising Knowledge in the Seventeenth Century

I will begin by establishing the seventeenth-century sense that the epistemological status of works is determined by their arrangement.[1] This idea enters the work of Boyle and his colleagues in the early Royal Society through the works of Francis Bacon. Bacon figured the production of secure natural knowledge as a staged process

---

[1] For a fuller discussion of these issues see Harriet Knight, *Organising Natural Knowledge in the Seventeenth Century: the Works of Robert Boyle* (PhD diss., University of London, 2003).

which begins in natural history, with the collection of individuated, observational data, and proceeds to give these general philosophical significance through textual manipulation. Apparently editorial decisions about arrangement thus become epistemologically significant. Bacon's new logic, which aims to produce formal natural philosophy, is essentially a process of organising data, and illustrates the crucial epistemological importance of the arrangement of particulars. Natural philosophy begins with the enumeration of instances: 'first of all we must prepare a *Natural and Experimental History*, sufficient and good; and this is the foundation of all'. However, simply accumulating data is not enough: 'natural and experimental history is so various and diffuse, that it confounds and distracts the understanding, unless it be ranged and presented to view in a suitable order. We must therefore form *Tables and Arrangements of Instances*, in such a method and order that the understanding may be able to deal with them.' The 'understanding', 'directed and guarded' by the rules of 'true and legitimate induction' may then form axioms, based upon this initial tabulation.[2]

Bacon's philosophically orientated true induction proceeds via further rounds of tabulation. 'The investigation of *Forms*' begins with the construction of a '*Table of Essence and Presence*', which presents instances in which the nature being investigated occurs. Next, a '*Table of Deviation, or of Absence in Proximity*', is needed to enumerate instances in which that nature is absent. A third table, '*of Degrees or Comparison*', records situations in which the discussed form increases or diminishes, to confirm that it is in direct proportion to the nature with which it is identified.[3] Once this 'Presentation of Instances to the Understanding' has been completed, induction continues by elimination, based upon further comparison of the tabulated data. 'Then indeed after the rejection and exclusion has been duly made, there will remain ... a Form affirmative, solid and true and well defined.'[4] In Bacon's new logic, then, processes which literally rearrange units of sense-data are crucial to the production of philosophical knowledge. Historical knowledge of individual instances is transformed into certain, philosophical knowledge of forms through the textual manipulation of data.

We can bring this discussion closer to Boyle by looking at the methodological prescriptions of his assistant, Robert Hooke. Hooke's 'General Scheme ... of the Present State of Natural Philosophy, and how its Defects may be Remedied' prescribes an initial collection of material to be 'laid up only in Heaps as it were, as in

---

[2] Spedding, Ellis and Heath, eds, *The Works of Francis Bacon*, London, 1857–74, IV, 127. On Bacon's tabulations see Paolo Rossi, *Francis Bacon: From Magic to Science*, London, 1968, 201–7, and Michel Malherbe, 'Bacon's Method of Science', in Markku Peltonen, ed., *The Cambridge Companion to Bacon*, Cambridge, 1996, 75–98.

[3] Spedding, Ellis and Heath, eds (1857–74), IV, 129, 137.

[4] *Ibid.*, 145–6.

a Granary or Store-House', which should 'from thence afterwards ... be transcribed, fitted, ordered and rang'd, and Tabled ... to be made fit for Use'.[5] Hooke devises a new technology for rearranging this data, but his end is the same as Bacon's: to render natural historical data philosophical through its organisation. His version of Bacon's repeated tabulations is an album, designed to accommodate 'Histories ... writ in brief, in a small piece of very fine Paper'. This will allow the collation of 'all the Histories belonging to any one Inquiry', and will crucially accommodate rearrangements: 'they may at any time, upon occasion, be presently remov'd or alter'd in their Position or Order'. The album is to be used to solve listed philosophical queries.[6] Here again particulars are self-sufficient units which gain significance through rearrangement.

I will now relate Boyle's comments on the disorderly form of his works to this staged model of knowledge creation. These suggest that his use of piecemeal forms reflects his sense that his works remain foundational to philosophy, only achieving the first historical data-heaping phase. Boyle's 'Proemial Essay' to his 1661 *Certain Physiological Essays* reveals his conviction that the content of works determines their proper form, and that for experimental work, which yields individuated and thus philosophically uncertain results, the essay form is appropriate, being evidently incomplete, piecemeal and unsystematic. Conventional 'Laws of Method' demand that natural philosophy is published in 'compleat Systems', Boyle claims, however, this formal requirement is 'none of the least impediments of the real advancement of true Natural Philosophy'. New discoveries are suppressed by their authors altogether, or overlooked in the midst of derivative works, whose 'specious and promising Titles and Comprehensive method' also suggest 'that all the parts of Natural Philosophy have been already sufficiently explicated', discouraging further research.[7] By contrast, as a diffident, piecemeal form, the essay is an ideal vehicle for presenting accounts of particular, individuated observations and experiments. 'Books of Essays', Boyle claims, offer the ideal form for natural knowledge.[8] During the 1670s, tract collections, with prefaces that emphasise their disjuncture and provisionality, duly became his characteristic mode of publication.[9]

Boyle's 'Advertisement ... about the Loss of many of his Writings', issued as a double-sided broadsheet in 1688, reveals that these attitudes endured throughout his

---

[5] Richard Waller, ed., *The Posthumous Works of Robert Hooke*, London, 1705, 21.

[6] *Ibid*., 64.

[7] Michael Hunter and Edward B. Davis, eds, *The Works Of Robert Boyle*, 14 vols, London, 1999–2000, II, 10–11.

[8] *Ibid*., 13.

[9] These include *Cosmical Qualities* (1670), *Rarefaction* (1671), *Flame and Air* (1672), *Essays of Effluviums* (1673), *Saltness of the Sea* (1673), *Hidden Qualities* (1674), and *Mechanical Origin of Qualities* (1675).

career. Cataloguing a series of thefts, losses, and damage to his papers, Boyle justifies the piecemeal publication of 'Remains and Fragments', as essential to their survival. This pragmatic defence of unsystematic publication is reinforced by invoking the nature of his subject:

> [as the] heaps of Fragments … mainly consist of … Matters of Fact; their being huddled together without Method … may not hinder them from being fit, if well dispos'd of, to have places some where or other in the History of Nature; and to become not Unserviceable Materials in the Structure that is aim'd at in this Age, of an usefull and well Grounded Philosophy.[10]

Readers of 'Experimental Philosophy' are presented as indifferent to 'the Method of Writings', 'at least after the first perusal', because they treat such texts as collections of particulars valuable for their potential to be rearranged. The 'Notions and Experiments themselves, abstracting from the Order they were deliver'd in' are 'the things that Philosophers use to take Notice of, and permanently retain in their Memories'.[11] Boyle expects his readers to retain items of content which they will redeploy in significant structures. His own arrangement is therefore largely irrelevant.

Criticism of Boyle's Piecemeal Forms

Throughout his career then, Boyle suggests that the piecemeal form of his works reflects the factual status of their content. As this contention is understandable in the light of Bacon and Hooke's explanations of knowledge creation, so it illuminates posthumous responses to Boyle's works, and especially the debate as to their intellectual scope, which has continued from the eighteenth century to the present day. While Boyle presents his use of piecemeal forms as a positive choice, commentators have repeatedly complained that Boyle failed to draw out the philosophical significance of his work, and that their disjointedness prevented others from doing so. For example, in the eighteenth century Boerhaave complained that Boyle's works 'are so many, and printed so separately, that it is exceedingly difficult to procure a compleat Collection'.[12] He was effusive about the range of Boyle's works, but saw their implications as having to be drawn out by readers: 'Which of Mr. Boyle's writings shall I recommend? All of them. To him we owe the secret of fire, air, water, animals, vegetables, fossils: so that from his works may be deduced the

---

[10] Spedding, Ellis and Heath, eds (1857–74), XI, 170–71.
[11] *Ibid.*, 170.
[12] Herman Boerhaave, *A New Method of Chemistry*, London, 1727, 47.

whole system of natural knowledge'.[13] John Freind's presentation of Boyle's works as representative of the state of early eighteenth-century chemistry similarly suggested that they provided a potential but unrealised basis for natural philosophy. Chemistry had been 'with great Industry improv'd and enlarg'd', but was not yet 'reduc'd to the Rules of true Philosophy', Freind claimed in his *Chymical Lectures*: he praised the 'laudable Progress in Experiments' but saw 'little Advances' towards 'Explication'. The significance of Boyle's work would have to be extracted from his actual legacy of collected natural historical data, Freind argued: 'he has left us plentiful Matter, from whence we may draw out a true Explication of things, but the Explication itself he has but very sparingly touch'd upon'.[14] Similar criticisms of Boyle as having failed to produce polished natural philosophy persist in nineteenth- and twentieth-century accounts.[15]

In the mid-twentieth century, Marie Boas Hall launched the classic attempt to reassert Boyle's philosophical importance. Her work presents a philosophical Boyle, 'the restorer of the mechanical philosophy' (a phrase she borrows from Peter Shaw's epitome): in brief, 'Boyle as a Theoretical Scientist'.[16] Significantly for my argument that organisation determines epistemological status, Boas Hall's project to present a coherent, philosophical Boyle is evidently strained when it encounters Boyle's texts: she repeatedly admits that her reading is complicated by the disorder of Boyle's writings. Her general project is to unify Boyle's corpus; to identify a principle which connects 'the seemingly formless mass of Boyle's philosophic writings' without recourse to the general influence of 'Bacon's emphasis upon experiment'. Positing the corpuscular philosophy as 'the connecting link' which gives intellectual coherence to Boyle's project makes his textual forms problematic, however.[17] The 'prolix, rambling, unmethodical form of his works' is 'paradoxical' in the light of his 'careful, systematic,

---

[13] Herman Boerhaave, *Methodus discendi medicinam*, quoted by Birch in 'The Life of the Honourable Robert Boyle', in Thomas Birch, ed., *The Works of the Honourable Robert Boyle*, 5 vols, London, 1744, I, 91.

[14] John Friend, *Chymical Lectures*, London, 1712, 1–4.

[15] For examples see William Brande, 'Chemical Philosophy', in the *Supplement to the Britannica*, Edinburgh, 1824, III, 15; A.M. Clerke on Boyle in Leslie Stephen, ed., *Dictionary of National Biography*, London, 1886, VI, 122; J.G. Growther, *Founders of British Science*, London, 1960, 51; D.C. Firth on Boyle in Donald Hutchings, ed., *Late Seventeenth-Century Scientists*, Oxford, 1969, 1–32.

[16] Marie Boas [Hall], *Robert Boyle and Seventeenth-Century Chemistry*, Cambridge, 1958, 1, and 'Boyle as a Theoretical Scientist', *Isis* (1950), **41**, 261. This emphasis on Boyle as philosopher is also apparent in A. Rupert Hall and Marie Boas Hall, *A Brief History of Science*, New York, 1964. Shaw calls Boyle 'the great restorer of mechanical philosophy among us', in Peter Shaw, ed., *The Philosophical Works of the Honourable Robert Boyle*, 3 vols, London, 1725, I, i. The arrangement of Shaw's edition aims to present Boyle as a philosopher, as I will show below.

[17] Boas [Hall] 1950, 264.

exact experimental technique', Boas Hall argues; formal shortcomings have directly, and deservedly, damaged Boyle's philosophical reputation.[18] For Boas Hall, then, the disorderly form of his works obscures Boyle the coherent thinker.

More recent accounts which present a philosophical Boyle operate similarly. For example, Rose-Mary Sargent's 1995 study of Boyle also reads his experimental work in the context of a '"comprehensive method" that would lead to knowledge in all areas of human concern', claiming that when his 'individual researches are taken together, it becomes clear that they were designed by him to contribute to one complex and coherent philosophical project'. Like Boas Hall, Sargent must also admit, however, that variety rather than coherence characterises his texts: 'Boyle's extreme eclecticism led him to publish works in many different areas, while his extreme caution led him to leave most of these studies in an unfinished state'. Boyle's importance 'from the standpoint of modern science' has been limited as a result of this textual incoherence: 'Boyle was not a great scientist in the tradition of Newton, because he never provided a systematic theoretical account for any of the areas that he studied'.[19]

For the same reason that scholars who emphasise Boyle's philosophy express dissatisfaction with his texts, those whose research focuses on those texts have tended to describe his work as factually orientated. Most famously, in *Leviathan and the Air-Pump* Steven Shapin and Simon Schaffer both attend to Boyle's 'literary technology', and present the establishment of 'matters of fact' as central to Boyle's intellectual project. Where Boas Hall attempts to understand Boyle's overall philosophical programme and laments the incoherence of his books as whole units, Shapin and Schaffer concentrate on the particularity of Boyle's works (for example, focusing on his presentation of accounts of individual experiments). In their reading, Boyle's disorderly, diffident presentation is chosen as suitable for establishing the credit of his experimentally validated matters of fact. While maintaining Boas Hall's account of Boyle as dedicated to the systematic testing of the corpuscular hypothesis required that the formlessness of his texts be dismissed as incidental, and referred to the vagaries of circumstances or character, for Shapin and Schaffer the textual presentation of information determines its reception as legitimate knowledge. Boyle's literary, material and social technologies directly govern the acceptance of his matters of fact.[20]

---

[18] Boas [Hall] 1958, 205. See also Boas Hall, 'Boyle's Method of Work: Promoting his Corpuscular Philosophy', *Notes and Records of the Royal Society* (1987), **41**, 112.

[19] Rose-Mary Sargent, *The Diffident Naturalist*, Chicago, 1995, 1, 205–6.

[20] Steven Shapin and Simon Schaffer, *Leviathan and the Air-Pump*, Princeton, 1985. Steven Shapin first identified Boyle's 'literary technology' in 'Pump and Circumstance: Robert Boyle's Literary Technology', *Social Studies of Science* (1984), **14**, 481–520. Shapin also emphasises Boyle's development of 'discursive means' for making experimental knowledge in *A Social History of Truth*, Chicago, 1994, 126–7.

Existing Boyle scholarship then suggests that 'literary' readings tend to posit a factual Boyle, while those which concentrate on his content are more likely to emphasise his adherence to the mechanical philosophy. Jan Golinski has noted this correlation, showing that treating chemistry as subordinate to philosophical theory 'has been associated, in the work of certain seminal historians of chemistry, with a failure to treat the linguistic and communicative practices by which chemistry was propagated and sustained'. He presents Boas Hall's work on Boyle as exemplary of the 'failure to focus on the language of chemistry as an object of historical scrutiny'.[21] I suggest that this phenomenon becomes understandable if we shift the focus of the 'literary' reading from language to form. Positing a coherently philosophical Boyle necessarily involves coming into conflict with the disorderly form of his published works. Not only in accounts like Boas Hall's, but in terms of the seventeenth-century understanding of the significance of organising knowledge, Boyle's disorderly works exist in tension with his philosophical aspirations. According to seventeenth-century epistemological theory, the same facts which constitute only historical knowledge when presented piecemeal and disordered can attain general, philosophical status through rearrangement. Formally, then, Boyle's most characteristic and distinctive texts suggest Boyle the natural historical collector rather than Boyle the systematising natural philosopher.

Eighteenth-Century Editions of Boyle

This sense of the importance of the arrangement of natural knowledge provides a new basis for assessing the eighteenth-century editions of Boyle's works. Robert Markley has suggested that in their 'rewriting of Boyle' these show 'how science is recontextualized and redefined ... into an autonomous, self-consciously disciplinary discourse'.[22] He claims that this redefinition operates through rephrasing. I contend, however, that while these editions do seek to redefine Boyle, they attempt this through the rearrangement, or methodisation of the collected works. This represents an attempt to apply a Baconian process of induction to the piecemeal, natural historically orientated collections of particulars in Boyle's corpus, thus advancing the status of his works from disorderly natural history to systematic natural philosophy. Boulton and Shaw's eighteenth-century editions should thus be seen in the light of complaints

---

[21] Jan V. Golinski, 'Chemistry in the Scientific Revolution: Problems of language and communication', in David C. Lindberg and Robert S. Westman, eds, *Reappraisals of the Scientific Revolution*, Cambridge, 1990, 368.

[22] Robert Markley, *Fallen Languages*, Ithaca, 1993, 217–18.

about Boyle's failure to draw out the philosophical implications of his work, as broadly Baconian attempts to achieve this through textual rearrangement.[23]

I will begin with a discussion of Richard Boulton's 1699 epitomised *The Works of the Honourable Robert Boyle*, which seeks to construct Boyle as a recognisably systematic philosopher by placing his works in a recognisably systematic framework. Boulton imposes a 'rational' arrangement on Boyle's works. There is a 'usual Method' for presenting '*Systems* of Philosophy', he claims: these 'begin with Generals, and thence ... proceed to Particulars; and ... lay down Principles in Order to Explain the several *Phænomena* of Nature by those Principles'. Boulton's epitome of Boyle is arranged 'in compliance' with these conventions: he presents 'Doctrin' and 'Principles', which are then 'Illustrated by Experiments and Observations'.[24] While Boyle's experimental essays re-enacted the process of discovery by narrating experiments and observations before proposing hypotheses based upon them, here experiments are presented as proofs of stated principles. This subordination of observations to principles is reinforced by Boulton's provision of a 44-page 'General Idea' of Boyle's works, which provides 'a short View' of their contents by presenting 'the most General Doctrines' to readers. The section contains no proofs, because experimental reports cannot be summarised. Readers who question its 'Doctrines' are referred to the main epitome for evidence. Similarly, Boulton's index 'only contains the General Heads of the Chapters'. However, these should be 'sufficient, since the Experiments which are contained in the Greatest Part are ranged under those General Heads, and designed as Proofs of them'.[25]

Boulton presents a philosophical Boyle who subordinates particulars to systematic ends, collecting instances 'in order to a general intelligible *System* of Philosophy'. He acknowledges that some of Boyle's works present 'several Historical Truths by way of Notes without Philosophizing upon them' designed 'to promote as well as perfect' knowledge, and he claims not to have interpolated connections here. Where Boyle has 'interspers'd' 'Philosophical Thoughts' with the 'Historical Truths', however, Boulton includes them, so 'that the Work might be more perfect'. Boulton's placement of Boyle within a 'philosophical' framework represents an attempt to 'perfect' his corpus. While Boulton has 'taken Care not to omit the least *Phænomenon* which might contribute

[23] Michael Hunter briefly discusses all of these editions in the 'General Introduction' to Hunter and Davis, eds (1999–2000), I, lxxxv–lxxxviii. J.R.R. Christie and J.V. Golinski suggest a similar reading of Shaw's edition in the context of highlighting the difficulties of applying natural philosophical theory to chemistry in the early modern period: 'The Spreading of the Word: New directions in the historiography of chemistry 1600–1800', *History of Science* (1982), **20**, 249.

[24] Richard Boulton, *The Works of the Honourable Robert Boyle, Esq. Epitomiz'd*, 4 vols, London, 1699, I, *B4r.

[25] *Ibid.*, IV, ²A2r–²A2v, A8r.

to the promoting of Natural Knowledge', the emphasis in his epitome is in drawing those phenomena into a natural philosophical system.[26]

Peter Shaw's 1725 'methodised' Boyle opens with the claim that Boyle's works have hitherto been 'vastly' underestimated because they consist of 'a miscellany of essays, upon a great variety of subjects'. While valuable material is latent in the works, few readers have 'been at the pains to collect, methodize, and regularly study' them. Their neglect provides the context for Shaw's 'just and regular abridgement', which aims to perform this methodising role, so that the potential of Boyle's works will become immediately evident. He has proceeded firstly by collating sections of material on particular topics, 'that they should together tend to compose one regular whole'. These have then been distributed under 'the respective general heads whereto they naturally belong'd'. This allegedly natural arrangement will simultaneously reveal the significance of individual works, and of the whole corpus, Shaw claims; it is a method 'whence the scope and design of each separate piece, might presently become conspicuous; and the sense and tendency of the whole, thus rang'd, be the better judg'd of'.[27]

Shaw claims to have arranged Boyle's particulars into a hierarchical, logical framework, 'that those which were introductory or fundamental, might stand first, and prepare the way for the rest; and that all might naturally follow, mutually illustrate, and, as much as possible, prove, confirm, and lead to each other'. The work follows a rational, progressive arrangement. As arranged by Shaw, Boyle:

> takes up his reader at the elements, or fundamental principles of things, and with exquisite judgment conducts him thro' all the regions of nature, to furnish him with objects, whereon to exercise his faculties; and, being first sollicitous to make him a general philosopher, leaves him prepared for any farther inquiry he shall think fit to make, into the works of nature or art.

In addition, as in Boulton, the reader is presented with a 'doctrine' ahead of the individuated natural historical evidence: Shaw has deemed it 'acceptable' to prefix 'some short account of the doctrine rang'd under' each of his 'distinct general heads', so 'that the reader might come a little prepared to the consideration thereof'.[28] While Boyle sought to avoid prejudicing readers by directing their interpretation of his recorded facts, Shaw explicitly prepares the reader to reach particular conclusions.

Shaw does make some concessions to 'historical' arrangement, providing a catalogue of Boyle's publications, with their dates, which provides a means of

---

[26] *Ibid.*, I, *B3r.
[27] Shaw (1725), I, i–iii.
[28] *Ibid.*, I, ii, iv.

reinserting a historical frame.[29] Moreover, he claims that the nature of the material on 'Pneumatics' made it unsuited for redistribution: he has been 'induced' to allow the air-pump experiments to 'stand in the order wherein he wrote them' and not to 'range them under several heads'. Chronologically presented, the results 'give the history of the air-pump, and its improvements from time to time': this would have been obscured by 'the other method', and that 'would prove a greater inconvenience, than the seeming disorder wherein they now occur'.[30] Shaw goes on to call for further research in the field: his 'disorderly' chronological arrangement here serves the same purpose as it does in Boyle, illustrating the incompletion of the subject and thus encouraging future collaboration. Shaw also admits that Boyle himself 'was very far from devising and making experiments, in order to confirm or establish a system, raised without the help of them', and 'draws no dogmatical conclusions from his premises'. He even goes so far as to suggest that Boyle's anti-systematic stance will prevent his works from ever being truly systematised: 'as Mr. *Boyle* never design'd to write a body of philosophy, ... 'tis not to be expected, that even the most exquisite arrangement, should ever reduce them to a methodical and uniform system, tho' they afford abundant materials for one'.[31] However, it is clear that for Shaw any such qualms can be countered by what he sees as the evident value of systematisation. His epitome makes Boyle an essentially systematic writer.[32]

Thomas Birch's 1744 *Works of Boyle* contrasts with the systematised collections of Boulton and Shaw. Birch emphasises the diversity of Boyle's works: 'the whole contains a very large collection of philosophical essays on a great variety of subjects', and presents them as individually valuable: 'the excellence of the several parts of them is universally acknowledged'. The aim of his collection is to increase the dissemination of these excellent parts, of which Hans Sloane had previously had the only 'complete set'. Birch does not suggest that the juxtaposition of texts will augment their value, except in terms of convenience.[33] While there is evidence that Birch and his collaborators initially planned to use a systematic arrangement based on Shaw's, in its published form Birch's collection is chronological: the 'most improved editions' of Boyle's works are 'disposed in the order of time, in which they were

[29] *Ibid.*, I, xxx–xxxi.

[30] *Ibid.*, II, 405.

[31] *Ibid.*, I, v–vi, iii. This tension in Shaw's edition is noted in Markley (1993), 245–6.

[32] Peter Shaw, ed.,*The Philosophical Works of Francis Bacon*, 3 vols, London, 1733, reveals a similar agenda, see esp. I, and iii, III, v. In a similar vein to my argument, Lisa Jardine and Alan Stewart discuss the significance of Spedding's nineteenth-century reordering of Francis Bacon's works in 'Editing a Hero of Modern Science', in Marina Frasca-Spada and Nick Jardine, eds, *Books and the Sciences in History*, Cambridge, 2000, 354–68.

[33] Birch (1744), I, A2r.

first published'.[34] This framework will reveal the 'rise, progress, and dependencies' of Boyle's discoveries, Birch explains, and conform with Boyle's 'own judgment': an authorisation supported with reference to Boyle's complaints about de Tournes' unauthorised Latin edition which obscured the order of publication and potentially complicated priority disputes.[35]

From the viewpoint of modern editorial theory, Birch's 1744 chronological arrangement of Boyle's full texts is far preferable to Shaw's abridged, methodised edition. Thus, while Michael Hunter acknowledges that Shaw's presentation of Boyle was 'in many respects ... more influential' than Birch's, he notes that it is 'open to many of the same criticisms as Boulton's' in its methodisation. Birch's edition is praised because: 'It presented the individual writings that it comprised largely in order of publication ... It was also complete, including all Boyle's theological as well as his scientific works. And, though it modernised the text, it did at last make Boyle's own words available in full'.[36] While accurate as an analysis of these collections by the standards of current scholarly editions, I contend that this reading may be complicated by a comparison of the divergent aims of the two collections. While Birch's edition is essentially engaged in a historical project of memorialisation, I would argue that Shaw sees his work as actively participating in Boyle's natural philosophical work, rather than as simply recording it. He includes footnotes relating to developments which have occurred since Boyle's time, and his edition was understood by an eighteenth-century commentator as the culmination of Boyle's work, and as changing its epistemological status through its methodisation. Richard Warren discusses Boyle's legacy and Shaw's influence on it in his *Oratio ex Harveii* (1769). Boyle, Warren argues: 'did not so much build a new chemistry on new foundations as purge the old one ... This man [Shaw] therefore arranged the *Boylean* farrago in an apt, distinct and orderly fashion ... he

[34] For Henry Miles' versions of the Shaw scheme see 'Miles to Millar', 18 August 1741, BL Add. Ms. 4229, 100r–101v. By 17 December 1741, C. Mortimer appears to take the chronological arrangement for granted, and it is not clear when or why the change occurred (BL Add. Ms. 4229, 122r–123v).

[35] On Boyle's dissatisfaction with Samuel de Tournes' Latin collected works (*Roberti Boyle ... Opera Varia*, Geneva, 1677), see Hunter, 'General Introduction', in Hunter and Davis, eds (1999–2000), I, lxxviii, and Adrian Johns, *The Nature of the Book*, Chicago, 1995, 508–10. Despite Birch's use of a chronological framework, John Hedley Brooke and Michael Hunter suggest that his failure to present the dates of publication of Boyle's works – a striking omission in the light of the preface – has contributed to a tendency by some scholars to treat Boyle's works as a 'static system' rather than recognising the evolution of his thought. Hunter and Brooke, foreword to R. Hooykaas, *Robert Boyle: a Study in Science and Belief*, Lanham, 1997, xvii.

[36] Hunter, 'General Introduction', in Hunter and Davis, eds (1999–2000), I, lxxxvi, lxxxv, lxxxvii–lxxxviii. For Shaw's influence on Birch see also Michael Hunter, *Robert Boyle by Himself and His Friends*, London, 1994, lvii–lviii.

at last laid down the principles of chemistry, so that now we may acknowledge this art to be truly philosophical'.[37]

My research provides the background necessary to understand this apparently alien analysis of the relative importance of Boyle and his editor. Given the seventeenth-century understanding of induction as the rational rearrangement of data, a careful, systematic editor can be credited with making a crucial epistemological input (turning 'farrago' into 'philosophy') in the very process of producing a methodised edition. Birch's 1744 edition has been regarded as the source of presentations of Boyle as simply empirical: I would argue that its non-interventionist, chronological arrangement, which represents the fragmentation of Boyle's works, is a crucial factor in this.[38] By contrast, Boulton's 1699 and Shaw's 1725 editions of Boyle's works present him as a systematic philosopher; their rearrangements seek to transform Boyle's natural historical facts into natural philosophical certainty. The crucial emphasis placed on the ordering of information in seventeenth-century epistemology illuminates the rationale and aspirations of these eighteenth-century methodised collections of Boyle's works. They participate in the seventeenth-century understanding of correct organisation as releasing the philosophical potential of collected, historical data. They thus present themselves, and were clearly understood by Warren, as having changed (and raised) the status of Boyle's works through their methodisation. They seek to rearrange seventeenth-century natural history into natural philosophy for the eighteenth century.

---

[37] Richard Warren, *Oratio ex Harveii*, London, 1769, 11–12. Original reads: Boyle 'qui tamen non tem nova chemiae extruxit fundamenta, quam dejecit vetera … Hic igitur, cujus desiderio omnes tenemur, farraginem Boyleanam apte, distincte, ordinate, disposuit … ea demum chemiae posuit principia, ut artem vere philosophicam esse jam tandem agnoscamus'. Many thanks to Stephen Clucas for assistance with the translation. F.W. Gibbs mentions Warren's assessment in 'Peter Shaw and the Revival of Chemistry', *Annals of Science* (1951), **7**, 224–5.

[38] Richard S. Westfall takes Birch's *Life* of Boyle as a crucial source of the conception of Boyle as empirical. Birch's refusal to rearrange Boyle's work into a more systematic frame than chronological order may reflect the same concerns as the biographical portrayal. Richard S. Westfall, 'Unpublished Boyle Papers Relating to Scientific Method', *Annals of Science* (1956), **12**, 63.

Chapter 4

Boundary Work: 'National Quarrels and Party Factions' in Eighteenth-Century British Botany

Susan McMahon

Natural history had become respectable by 1700. It was also an enterprise practiced by an identifiable, disciplined community of natural historians who had as their central aim the first-hand observation of God's creation and its systematic organisation. Responsible natural history was a deliberate and self-conscious scientific undertaking, erected upon a foundation of careful observation, reliable witnesses and valid experiential techniques. The discipline involved sound philosophical principles, the consensual establishment of matters of fact, the deployment of appropriate social strategies, and the development of instrumental and methodological technologies. In particular, methods of classification had become entrenched as powerful analytical and predictive tools, and had begun to transform the practices of natural history in the early eighteenth century.[1] Nowhere is the methodological transition to classification more apparent than in the field of botany, which also enjoyed the distinction and high prestige of big science.[2]

Many in this great age of systems were soon preoccupied with inventing novel ways of ordering nature. However, as early as 1708, an aspiring botanist, Patrick Blair MD of Dundee Scotland, had become concerned that the efflorescence of new methods of botanical classification was both unnecessary and confusing. He worried

---

[1] Paul Lawrence Farber, *Finding Order in Nature: the Naturalist tradition from Linnaeus to E.O. Wilson*, Baltimore and London, 2000; Phillip Sloan, 'John Locke, John Ray and the Problem of the Natural System', *Journal of the History of Biology* **5(1)**, 1972, 1–53; Michel Foucault, *'The Order of Things': An Archaeology of the Human Sciences*, Gallimard edn, London, 1977.

[2] Mordechai Feingold, 'Mathematicians and Naturalists: Sir Isaac Newton and the Royal Society', in Jed Z. Buckwald and I. Bernard Cohen, eds, *Isaac Newton's Natural Philosophy*, Cambridge, MA and London, 2001, 77–102; Paula Findlen, *Possessing Nature: Museums, Collecting and Scientific Culture in Early Modern Italy*, Berkeley, 1994; Harold J. Cook, 'The Cutting Edge of a Revolution: Medicine and natural history near the shores of the North Sea', in J.V. Field and Frank A.J.L. James, eds, *Renaissance and Revolution: Humanists, Scholars, Craftsmen and Natural Philosophers in Early Modern Europe*, Cambridge, 1993, 45–72.

that the existence of many different methods would make it more difficult to become competent as a botanist. What was more important for Blair, however, was that the diversity of different methods had the potential to foster disagreement within the natural history community and to disturb the harmony and collegiality of the enterprise. He believed that the array of new methods would ultimately result in an environment where 'heresies and schisms may come to raise so many heats in our more innocent study of botany as they do already too much in the more worthy matters of religion'.[3] Blair's self-appointed mission was to reconcile what he saw as the contradictions inherent in the differing botanical classifications. His primary strategy was a plan to convince the botanical community to abandon the generally accepted method for classifying plants elaborated by John Ray, FRS (1627–1705), which had dominated English botany since the late seventeenth century, and to replace Raian taxonomy with a synthetic method of Blair's own devising. The consequence of his ambitions was rather to unite the '*Totus Botanicorum Modernorum chorus*' in their defence of the established scientific culture and in their unanimous rejection of Blair's heterodox proposals for interpreting nature.[4] This paper explores this largely invisible episode of early eighteenth-century botany by analysing how the entrenched community of natural historians responded to Blair's challenge. Fundamental to my analysis is the claim that natural history was a contingent social and cultural enterprise of time and place. Early modern natural history was not simply a matter of obviously justified beliefs, and it was indeed possible to claim that there were many equally valid and objective ways of classifying nature. In this context, there was no necessarily rational basis for the cultural dominance of any unique model to govern natural history and its practices.

I have attempted to understand how the discipline of natural history was constructed in early modern England and how its content, practices and standards were maintained within a specific natural history community. Inevitably, my research has drawn upon the sociological implications of Thomas Kuhn's *The Structure of Scientific Revolutions* (1970, 2nd edition 1989) that normal science employs standardised and accepted procedures, largely assumes the correctness of existing knowledge and,

[3] British Library, Sloane Manuscripts 3321, letter from Patrick Blair to James Petiver, Dundee 30 November 1708, ff. 237–8, hereafter Sloane MS.

[4] The process of demarcating orthodox science and its practices is called boundary work by sociologists. See for instance Thomas F. Gieryn, *Cultural Boundaries of Science: Credibility on the Line*, Chicago and London, 1995; Gieryn, 'Distancing Science from Religion in Seventeenth-Century England', *Isis*, **79**, 1988, 582–93; Steven Shapin and Simon Schaffer, *Leviathan and the Air Pump*, Princeton, 1985; and Shapin, 'The Politics of Observation: Cerebral anatomy and social interests in the Edinburgh phrenology disputes', in Roy Wallis, ed., *On the Margins of Science: the Social construction of rejected knowledge*, Sociological Review Monograph 27, Keele, 1979, 139–78.

especially, that scientific knowledge itself incorporates the beliefs and values as well as the cognitive commitments of a particular community. What achieves acceptance as orthodox scientific knowledge does so by processes acquired through socialisation and maintained by the application of authority and mechanisms of social control.[5] Thus, orthodox or normal science is located in the social practices, values and assumptions of the community, and is not merely the determination of 'true matters of fact' or the application of unambiguous rationality. Rather, the scientific community itself is actively involved with constructing and policing the intellectual and social boundaries of their enterprise according to a set of fluid and historically contingent standards.[6] What is important about Blair's story is that the stabilisation of a specific tradition of natural history in the early eighteenth century, especially with respect to the adoption of a particular method of biological classification, hinged upon the consent of a self-defined community of active and interested natural historians who were located within the localised scientific culture of early modern England.[7] The community agreed upon the standards of the discipline, evaluated knowledge claims and made the judgements as to who would be counted as a creditable member.

At least since the 1680s, the natural historical project had centred upon John Ray (1627–1705), a fellow of the Royal Society and author of the paradigmatic work on natural theology, *The Wisdom of God Manifested in the Works of Creation* (London, 1691). Ray's stature as a natural historian, and especially as the 'Great Botanist of our age' was unrivalled. His monumental three-volume *Historia Plantarum* (1686, 1688 and 1703) was the principal botanical text in Britain and the *Joannis Raius Synopsis Methodica Britannicarum* (1690 and 1696) the standard work of plant taxonomy. Ray's classification method, which distinguished plants according to an array of specific differences, had been adopted by medical professionals to organise *materia medica* and was also in use among a wider English community of *botanophili*, a learned botanical and horticultural interest, which flourished among the wealthy landowning class.[8]

---

[5] Thomas Kuhn, *The Structure of Scientific Revolutions*, 2nd edn, Chicago, 1989, 176–87. See also Barry Barnes, *T.S. Kuhn and Social Science*, London, 1982.

[6] Thomas F. Gieryn, 'Boundary-work and the Demarcation of Science from Non-Science: Strains and interests in professional ideologies of scientists', *American Sociological Review*, **48**, 1983, 781–95; Brian Wynne, 'Between Orthodoxy and Oblivion: the Normalization of deviance in science', in Wallis, 1979, 67–84; and Wynne, 'C.G. Barkla and the J. Phenomenon: a Case study in the treatment of deviance in physics', *Social Studies of Science*, **6**, 1976, 307–47.

[7] There is an array of work in the sociological literature about the invention of scientific tradition. For an introduction, see Robert K. Merton, *On Social Structure and Science*, Chicago, 1996. See also Eric Hobsbawm, *The Invention of Tradition*, Cambridge, 1992.

[8] Douglas Chambers, 'Stories of Plants: The assembling of Mary Capel Somerset's botanical collection at Badminton', *The History of Collections*, **9(1)**, 1997, 49–60; and Dawson Turner, 'Introduction', *Extracts from the Literary and Scientific Correspondence of R. Richardson, of Bierly, Yorkshire: Illustrative of the state and progress of botany*, London, 1835, xi.

Ray's continuing posthumous authority was such that, as early as 1708, at least one author who claimed to have improved upon Ray's method for classifying plants was unable to find a publisher for the work.[9] When an English translation of Joseph Pitton de Tournefort's *Botanical Institutions* appeared in 1719, it was thoroughly Anglicised. Although deploying Tournefort's scheme for classifying plants, the edition utilised Raian nomenclature, terminology and descriptive practices for its British audience.[10] In short, Raian natural history had become fully entrenched in the English community, and Ray himself was portrayed as the sanctioned, socially acceptable 'father' of English botany.

Ray had played a commanding role in building a coalition of natural historians composed of independent gentlemen, clergymen, physicians and fellows of the Royal Society. As early as 1690, there were at least 60 individuals directly involved with and contributing to Ray's natural history project in England, and in subsequent years these numbers continued to increase as new members were recruited to this highly respectable undertaking.[11] One of Ray's greatest allies was Hans Sloane, Secretary of the Royal Society (1693–1700) and a member of Council (1690–99). As editor of the Royal Society's *Philosophical Transactions* (1693–1712), he was also in a unique position to advance natural history and the interests of the natural historians.[12] Sloane, who had been praised by Ray as an 'a man most learned and extraordinarily skilled in res herbaria',[13] would subsequently become prominent within scientific circles more generally as a foreign member of the Academie Royale des Sciences (1708), vice-president of the Royal Society (1712), president of the Royal College of Physicians (1719–35) and, finally, president of the Royal Society (1727–40), in addition to the influence he possessed through his appointments as physician to Queen Anne and

---

9 *Correspondence of R. Richardson* (1835), letter from Edward Lhwyd to Dr Richard Richardson, October 8, 1708, 95. 'Mr. Buddle hath drawn up a new *Synopsis Plant. Brit.*; but he doubts whether he can get it printed; tho he supposes it a very considerable improvement of Mr. Ray's, who, he says, wanted many things to compleat his. He adds that he improves the method by the help of Tournefort, Rivinus, &c. and that he often refers to figures and corrects vicious ones'.

10 *The Compleat Herbal: or, the Botanical Institutions of Mr. Tournefort, chief Botanist to the late French King. Carefully translated from the Original Latin. With large additions from Ray, Gerarde, Parkinson, and others, the most celebrated Moderns*, London, 1719, where 'the descriptions [were] generally borrowed from Ray, whose Accuracy and Diligence have procured him a merited Reputation'. Quotation taken from publisher's preface on page 3.

11 Susan McMahon, *Constructing Natural History in Britain (1650–1700)* (PhD diss., University of Alberta, 2001).

12 For Sloane's editorship of the *Philosophical Transactions*, see T. Christopher Bond, 'Keeping Up with the Latest Transactions: the Literary critique of scientific writing in the Hans Sloane years', *Eighteenth Century Life*, **22(2)**, 1988, 1–17.

13 *Erudissimus Vir & rei herbaria pertissimus*. John Ray, *Historia Plantarum* (1686), 228.

later to George II. Sloane became known as 'the great patron of natural learning',[14] and subsequent generations of natural historians courted his favour as the person in whose hands 'all things centre' in affairs of natural history.[15]

By the 1720s, the botanical community consisted of an array of individuals who pursued their own interests, ambitions and intellectual commitments, but who were also united by a genuine concern with *res herbaria* and with the responsible prosecution of their discipline. In addition to Hans Sloane, members of this loose collective included, among others: Tancred Robinson FRS, Ray's '*alpha amicorum*' and physician to the king; the Yorkshire physician Richard Richardson FRS; the clergyman John Morton FRS, author of the *Natural History of Northamptonshire*; and Samuel Dale of Essex who had been Ray's personal physician. Another important member of the network was William Sherard FRS, former consul at Smyrna (1703–16) and friend of Herman Boerhaave (1668–1738), professor of Medicine at Leiden University. There was also a younger generation of botanists eager to make their reputations in the discipline, such as John Martyn (FRS 1727), founder of the Botanical Society of London and future professor of Botany at Cambridge, and Mark Catesby (FRS 1733) a client of Sloane and Sherard and author of the *Natural History of Carolina, Florida and the Bahamas Islands* (1726).[16]

Patrick Blair, an ambitious physician from Scotland, would follow the customary means of achieving scientific credit within the informal natural history collective. He was one of many who sought Sloane's sponsorship, and his correspondence with Sloane is entirely characteristic of what we have come to expect of early modern client–patron relationships, with professions of friendship, expressions of obligation, and tendering of gifts.[17] Blair had begun writing to Sloane as early as 1705,[18] and soon his observations were being reported in the Royal Society's *Philosophical*

---

[14] Sloane MS 4047, ff. 40.

[15] *Correspondence of R. Richardson* (1835), letter from William Sherard to Richardson, 6 April 1723, 194.

[16] For biographical information, see Ray Desmond, *Dictionary of British and Irish Botanists and Horticulturalists including Plant Collectors, Flower Painters and Garden Designers*, London, rev. 1994; D.E. Allen, 'John Martyn's Botanical Society: a Biographical Analysis of the Membership', *Proceedings of the Botanical Society of the British Isles*, **6**, 1967, 305–24; and the *Dictionary of National Biography*, hereon *DNB*.

[17] Findlen, 'Patrons, Brokers and Strategies' (1994), 346–93; Lesley B. Cormack, '"Twisting the Lion's Tail": Practice and theory at the court of Henry, Prince of Wales', in Bruce T. Moran, ed., *Patronage and Institutions: Science, technology and medicine at the European Court 1500–1750*, Rochester, 1991, 67–83; Linda Levy Peck, *Court, Patronage and Corruption in Early Stuart England*, London, 1990; Mario Biagioli, 'Galileo's System of Patronage', *History of Science*, **28**, 1990, 1–62.

[18] DNB, II, 623; Richard Pulteney, *Sketches of the Progress of Botany in England from its Origin to the Introduction of the Linnaean System*, II, London, 1790, 134–40.

Transactions, followed by his election as a fellow of the Royal Society in 1712.[19] Blair's next move was to publish the *Miscellaneous Observations in the Practice of Physick, Anatomy and Surgery. With new and curious remarks in botany* (London, 1718), which also contained several of his letters to James Petiver FRS, demonstrator of plants at the Apothecaries' Physic Garden at Chelsea. In what was surely intended to be a tactic for self-advancement, Blair dedicated his *Botanick Essays* (London, 1720) to Sir Isaac Newton in his capacity as president of the Royal Society, and Sir Hans Sloane as president of the Royal College of Physicians.[20]

The five separate essays composing the *Botanick Essays* were primarily concerned with taxonomy, physiology and plant sexuality, the central concerns in contemporary botany. Blair's declared ambition was to advance the 'progress' of botany and he intended to achieve this by a process of reconciliation and recommendation. His strategy was to provide his personal assessment of the variety of opinions existing in each domain, and then proceed 'to add what I think has been wanting, to correct what by proper Experience I find they have advanced amiss, and to make several Discoveries and Improvements upon the whole'. In fulfilling this goal, Blair also declared himself to 'have behav'd so impartially, that I have given a full View of the Advantages, nor have I expose'd the Imperfections of any, beyond what was necessary to clear up the Truth'.[21] It is true that the *Botanick Essays* provides valuable insight into the particular concerns and controversies that animated the discipline in the early eighteenth century. However, Blair was seldom tactful in expressing his opposition to 'novel Opinions, which only serve to pervert, and be a Mask to disguise the Truth'.[22] What is more important, many of Blair's own opinions were deviating from the views held more generally within the English botanical community.

One of the fundamental questions about plants in the late seventeenth century had been whether they possessed sex and reproduced sexually. Nehemiah Grew and John Ray had both been of the opinion that plants were sexual and Ray especially had argued

---

[19] Patrick Blair, 'Osteographia Elephantina: Or a full and exact description of all the bones of an elephant, which died near Dundee, April the 27th, 1706', *Philosophical Transactions* **27(326)**, 1710–12, 53–116; Blair, 'A Continuation of the Osteographia Elephantina: Or, a description of the bones of an elephant, which died Near Dundee, April 27th, 1706', *Philosophical Transactions* **27(327)**, 1710–12, 117–67; and Blair, 'Part of a Letter from Mr. Patrick Blair to Dr. Hans Sloane R.S. Secr. Giving an Account of the Asbestos, or Lapis Amiantus, Found in the High-Lands of Scotland', *Philosophical Transactions* **27(333)**, 1710–12,434–6.

[20] Patrick Blair, 'Dedication', *Botanic Essays* (1720). Blair continues that 'I was once afraid that the agreeable Science of Botany should be at a Loss by the Death of Mr. Ray and several of his Correspondents, but I am glad to find that it still continues in its former Vigour, under the happy Influence of Sir Hans Sloane, President of the College of Physicians, to whom I have been singularly obliged'. Blair, 'Preface' (1720), Sig. a2v.

[21] Blair, 'Preface' (1720), n.p.

[22] *Ibid.*, 325.

that pollen played a role in plant reproduction; Rudolph Jacob Camerarius (1665–1721) at Tübingen later confirmed the function of pollen.[23] Blair, however, was out of step with the more generally accepted mechanical interpretations of sexual fertilisation, and argued instead that an effluvia or immaterial seminal principle carried within the sap of the male plant was the means by which the female was fertilised, analogous as to how Blair saw impregnation occurring in animals.[24] Responding directly to Blair, Richard Bradley FRS (1688–1732), professor of Botany at Cambridge and a popular and prolific writer on botanical and horticultural matters, undertook a series of controlled experiments on plant sexuality by removing the pollen-producing stamens of the flowers in an isolated population of plants. He concluded that because the plants had produced no viable seeds, pollen must have a role in fertilisation. In his response to Blair, Bradley argued for a more conventional mechanistic explanation and he hypothesised a physical, albeit unobserved, action of the pollen in plant fertilisation.[25] In return, Blair sought allies in the affair against Bradley[26] and attempted to court Sloane's support by offering to dedicate to him an entirely new monograph on generation. In the event, however, Blair was unable to garner sufficient backing for his views on the *semen insubstantia* and this work was never printed.[27] Furthermore, his disagreements with Bradley were extremely acrimonious; although his public

---

[23] A.G. Morton, *History of Botanical Sciences: an Account of the development of botany from ancient times to the present day*, London, 1981, 213–20. Grew's work on plant physiology enjoys the distinction of being the first research project directly sponsored by the Royal Society. See Michael Hunter, 'Early Problems in Professionalizing Scientific Research: Nehemiah Grew (1641–1712) and the Royal Society, with an unpublished letter to Henry Oldenburg', *Establishing the New Science: the Experience of the early Royal Society*, Woodbridge, 1989, 261–78.

[24] Blair, 'Essay IV' (1720), 222–326; Blair, 'Observations upon the Generation of Plants, in a Letter to Sir Hans Sloane, Bart. Pr. Coll. Med.', *Philosophical Transactions* **31**, 1720–21, 216–2l; Philip Ritterbush, *Overtures to Biology: the Speculations of eighteenth century naturalists*, New Haven, 1964, 88–108.

[25] Richard Bradley, *New Experiments and Observations, Relating to the Generation of Plants: Occasion'd by a letter lately publish'd in the Philosophical Transactions by Patrick Blair, MD, FRS. Together with an account of the extraordinary vegetation of peaches, abricots, nectarines ... as they were artificially cultivated this spring*, London, 1724.

[26] Bodleian Library, Rawlinson Manuscripts 323, letter from Patrick Blair to Mr Paul Bouiser, Surgeon, Boston, June 23, 1723, hereafter Rawl. MS 323. See also Rawl. MS 323, letter from Blair to an unidentified 'Worthy and kind sir', Boston, 3 or 5 December 1724, where he again discusses whether 'the *semen masculinum insubstantia* enters the uterus'.

[27] Martyn also advised Blair that 'I am afraid Sir Hans will not be pleased with being made a party in the dispute. His publishing your letter ['Observations upon the Generation of Plants'] was not giving his assent to the doctrine maintained in it. He was willing the world should know what you had to offer & if any objected against it, it could be no offence to him'. Rawl. MS 323, letter from Martyn to Blair, London, 23 June 1724.

utterances may fall within the limits of conventional civility, privately Blair's attack on Bradley was abusive.[28]

In another highly speculative essay on plant physiology from the *Botanick Essays*, Blair was also convinced that he had proven that sap circulated uninterruptedly throughout the plant. He argued that sap moved upward through specific dedicated vessels and downward through another set of dedicated vessels both of which were connected by capillaries, and in a manner similar to the circulation of blood in animals.[29] While it is not uncommon to find comparisons between plant and animal circulation in contemporary literature, the pioneering work of Grew in the 1670s had failed to find any evidence to support a hypothesis for vegetable circulatory systems. Ray and others had been in agreement, although a commonly accepted explanation describing the movement of sap was still in the future. By 1724, Blair could complain that his essay on sap had 'become so odious to some that they disdain to read it'.[30] In 1725, the experimental results of Stephen Hales FRS (1677–1761) on plant physiology were read before the Royal Society, and were seen to decisively disprove the notion of plant circulation.[31] Later published as the *Vegetable Staticks* (London, 1727), Hales' important work has been considered a model of the application of Newton's mixed mathematical methods to the study of plants, which set the experimental standard for future research on plant physiology.

The responses by Bradley and Hales successfully deployed experimental evidence to uphold more conventional interpretations of plant function as well as to counter Blair's atypical opinions on the vegetable economy; indeed, the greater evidential weight given to experimental methods seem to have rendered Blair's views untenable. However, Blair's objections to botanical classification methodologies were much more problematic, especially given that there is no single and unambiguously true method of classifying nature and hence no self-evidently correct rationale for the primacy of any particular method in the early eighteenth century. Nevertheless, by 1720 most natural historians agreed that their enterprise demanded the systematic collection and orderly digestion of creation; and, further, that the best methods of classification would offer the most effective means to acquire new knowledge. What the natural historians did not agree upon, however, was which method was unambiguously the

---

[28] Martyn's diplomatic editing ensured that most of Blair's vitriol did not make it into print. See, for instance, Rawl. MS 323, letters from Martyn to Blair, 23 June 1724 and July 7 1724. Nevertheless, Bradley's negative posthumous reputation owes much to Martyn who succeeded him as professor of Botany at Cambridge in 1732.

[29] Blair, 'Essay V' (1720), 331–408.

[30] Rawl. MS 323, letter from Blair to Martyn, Boston, 29 November 1724.

[31] Stephen Hales, *Vegetable Staticks: Or, an account of some statical experiments on the sap in vegetables*, London, 1727; see also Henry Guerlac, 'Stephen Hales' in C.G. Gillispie, ed., *Dictionary of Scientific Biography*, New York, 1972, 35–48; and Ritterbush, 1964, 57–88.

'best'. Moreover, the method that was seen to be 'best' in the conditions of the early eighteenth century often depended on an array of local circumstances, which could include pedagogical tradition and inherited collective judgement, but also entirely mundane issues such as the availability of particular plants to study and the intended uses of the classification itself. In England, of course, Ray's scheme was predominant among dedicated botanists, and even Blair agreed 'Ray's method being that which is more generallie received, especially in Great Britain'.[32] But the French preferred the easier classification by Tournefort, the Germans used the method of Rivinus and it is safe to assume that other methods similarly gained acceptance in other local circumstances. For instance, in the Netherlands Herman Boerhaave recommended that the methods of both Ray and Tournefort could be used in concert so that one would become 'more perfect in the knowledge of plants': Tournefort's because it was easier to use to identify species and Ray's because it was best for knowing the genera.[33] Thus, in practice, there was little difficulty with the existence of an array of incompatible methods, since only one method of classification, or two in the case at Leiden, tended to function within any local milieu.

Blair, however, had long feared that the increasing number of classification methods presaged 'a growing evil which cannot fail to ensue if every one continue at this rate to advance a new [method] of his own' and which would furthermore disrupt the 'excellent harmony' of the science. By 1712, he was already proposing that the aim of 'every serious Botanist' should be to find a compromise solution by 'reconciling the different methods'.[34] In the *Botanick Essays* of 1720, Blair described a history of botany where numerous and equally valid classifications were not only in competition with each other but had 'encreased [sic] faction and division' in the science and indeed had 'actually broke out into a Paper and Botanick War' between Ray and Tournefort during the 1690s.[35] However, in highlighting the Ray–Tournefort affair, Blair was also disregarding the practical accommodation that the respective local communities had achieved with the closure of that debate and where the acceptable and proper methods among the local English and French botanists had maintained a distinctly nationalistic profile.[36] Therefore, Blair fabricated a problem where the multiplicity of methods were seen to have already resulted in confusion and disunity within the discipline and he predicted this would serve to 'render Plants as unintelligible by Method, because of their great plurality, as formerly it was to know

---

[32] Sloane MS 3321, letter from Blair to Petiver, Cowpar of Angus, 24 March 1712, ff. 275–6.

[33] Herman Boerhaave, *A Method of Studying Physick*, translated into English by Mr Samber, London, 1719, 130–32.

[34] Sloane MS 3321, letter from Blair to Petiver, Cowpar of Angus, 27 August 1712, ff. 283v–4.

[35] Blair, 'Preface' (1720), n.p.

[36] Susan McMahon, 'John Ray, Joseph Tournefort and Essences: The "species" problem in the seventeenth century', *Journal of the History of Biology*, forthcoming; and Sloan (1972).

the plants without Method'.[37] His initial solution: botanists should stop inventing new methods and instead 'impartially correct and amend what they find amiss in former methods'.[38] Secondly, botanists should become competent in *all* the different classifications 'so whoever shall be at Pains to observe all the several Methods, they may come thereby to know more easily the several parts of the Plants' to properly identify and classify them.[39] We may consider Blair's intentions reasonable and his ambitions laudable. However, Blair continued his 'impartial' essay on contemporary taxonomic methods by providing an extended vindication of his fellow Scotsman, Robert Morison (1620–84), Physician to Charles II and author of *Plantarum Historiae Universalis Oxoniensis* (Oxford, 1682), in which Morison had proposed a classification not acceptable to the English botanical community. Blair's apologia was accompanied by a censure of John Ray who, in Blair's account, was wholly responsible for Morison's lack of recognition both as a botanist and as the 'restorer' of method, but especially for Morison's reputation as having illegitimately 'borrowed' his intellectual ideas from the Renaissance botanist Andreas Cesalpino.[40] Later, Blair would become concerned with trying to displace the culturally dominant Raian classification for organising plants, and to replace it with his own synthetic system. By that time, Blair himself would be the cause of another 'paper and botanick war' that erupted within the discipline.

In 1723, Blair presented Hans Sloane with a manuscript of the first two decades of another project on plants, the *Pharmaco-Botanologia*. Blair had been hoping to dedicate it to Sloane and thereby receive his protection.[41] The first instalment, however, was printed without a dedication and thus without Sloane's seal of approval. The *Pharmaco-Botanologia* was intended to be the practical application of Blair's proposal that botanists know all the different classifications. At the same time, he was suggesting that it was possible to know the medicinal virtues of plants by their physical appearance. Blair believed, optimistically, that all the existing incompatible and conflicting methods could be synthesised into one harmonious meta-classification that would enable the identification of individual plants and, as well, lead to the discovery of their medical virtues.[42] In Blair's words:

---

[37] Blair (1720), 72.

[38] *Ibid.*, 73.

[39] *Ibid.*, 219.

[40] John Ray, 'Preface', *Joannis Raii de variis plantarum methodis dissertio brevis in qua agitur*, London, 1682; Joseph Pitton de Tournefort, *Elemens de botanique, ou méthode de connoître les Plantes*, Paris, 1694, 19; and Blair (1920), 89–114.

[41] Sloane MS 4047, letter from Blair to Sir Hans Sloane, Boston, January 1723, ff. 116.

[42] Patrick Blair, *Pharmaco-Botanologia, or an Alphabetical and Classical Dissertation on all the British Indigenous and Garden Plants of the New London Dispensatory*, London, 1723–28, v. See also Blair, 'A Discourse Concerning a Method of Discovering the Virtues of Plants by their External Structure', *Philosophical Transactions* **31**, 1720–21, 30–38.

> By making the Alphabet my rule, and the classical [that is, according to medicinal virtues] Disposition my Method, I have framed such a Botanical Synopsis, as may afford Examples of the several Tribes of European Plants, and give an Inlet into the several Methods into which they have been reduced. I have shown how far their virtues suit with the Characters, reconciled the seeming Discrepancy among the several Methods, and order Matters so, that by looking into the Tribe you may guess at the Virtues. You may know the Tribe by its Characters, and distinguish each individual Plant or species by its peculiar Notes; by which an agreeable Harmony may appear in their Classes, Sections, Genera, Species, Characteristick and Distinctive Notes, Virtues and Uses.[43]

Blair's plan in the *Pharmaco-Botanologia* was neither to be confined to any one method of classification, nor to order plants either alphabetically or by medical virtue. Rather, his plans seem to have been to informally group similar or related plants together according to undefined notions of 'similar' or 'related', thereby de-emphasising the differences between plant species. His actual practice was to include a discourse on related species as their relatives were introduced according to the alphabetical ordering.[44] To try to clarify Blair's method, his first example was that of the genus *Abies*, or the fir, under which he described several species of fir trees. To this he joined an extended discourse on another genus of coniferous trees, that of *Pinus*, or the pine, again with a discussion of several species. Many 'curious and useful remarks' were included about medicinal or utilitarian uses and instructions for the propagation and culture of *Abies* and *Pinus* as well as other evergreens were discussed. Blair's rationale for joining these two genera of trees was that they were both 'of the same Tribe or Family', that they were 'big trees', evergreen, possessed amentaceous, or catkin-like flowers and had conical scaly hard fruit.[45] Today we also see *Abies* and *Pinus* as close relatives, constituting two of several genera of the family *Pinaceae.* It is even likely that we would be able to distinguish *Abies* from *Pinus* on the basis of Blair's rather unsystematic physical description. However, the organisation of the *Pharmaco-Botanologia* meant that finding reliable information on any individual plant was not entirely straightforward, although such identification may have been aided by using a companion such as *A New Table of Dispensatory Plants Distributed According to their Virtues* which was also available through the booksellers.[46]

Blair informs us that he adopted a mixed approach for organising plants in the *Pharmaco-Botanologia* because 'Mr. Dale had prevented me in the Prosecution of Mr.

---

43 Blair, 'Preface' (1723–28), iii–iv.

44 *Ibid.*, 'Preface', 'Decade II'.

45 *Ibid.*, 'Decade I', 1–8.

46 Patrick Blair, *A New Table of Dispensatory Plants Distributed According to their Virtues*, London, 1722. 'Sold by W. & I. Innys at the West end of St. Paul's, separately, or together with Dr. Blair's Botanick Essays'.

Ray's Method'.[47] However, the work does not appear to have been designed to attract an intended audience of apothecaries or other respectable medical professionals.[48] Blair's pedagogical method had been to comment on the plants contained in the *New Dispensatory.*[49] Despite popular and widespread use of alternate pharmacopoeias, the London College of Physicians traditionally dictated the official plants that could be used by apothecaries in medicinal preparations, licensed the printing of the official pharmacopoeia and had historically resisted the use of unauthorised versions. Traditionally, *materia medica* was listed alphabetically, but, since 1693, Ray's botanical arrangement had also been used for the 'officinal plants', a project which had been undertaken by Ray's close friend and neighbour, Samuel Dale, MD. The censors of the college had licensed a *sumptibus* edition of Dale's dispensatory, the *Pharmacologia* (London, 1693), subsequently reissued in new editions until 1739.[50] Thus, the *Pharmacologia* was an official publication, legally sanctioned by the London College of Physicians and endowed with its authority. Further, the work was printed under the auspices of the Royal Society throughout its several editions, and so Dale's *Pharmacologia* may best be seen as a cooperative venture between the London College of Physicians, the Guild of Apothecaries and the Royal Society; prominent members of all three institutions contributed to the project and endorsed its authority. Blair's proposals could provide no advantage over the familiar alphabetical listing for *materia medica*, nor did it improve upon Ray's taxonomy officially represented by Dale's work.

Blair's enterprise, however, was also unlikely to attract an audience of dedicated botanists whose agreed-upon undertaking was that of identifying, describing and classifying plants, and who were the primary audience for scholarly works on the topic. In fact, Blair complained that market was so 'glutted' with Ray's botanical works in Latin that 'nothing on that Subject and that Language' would be acceptable

---

[47] Blair, 'Preface' (1720), iv.

[48] Blair aimed at a botanically literate readership rather than targeting a non-specialist audience, although it is true that an appetite for popular pharmacological works may have been inexhaustible in the period. For comment on popular pharmacopoeias, see David L. Cowen, *Pharmacopoeias and Related Literature in Britain and America 1618–1847*, Aldershot, 2001.

[49] Blair, 'Preface' (1723–28), iii–iv. Blair did not specify an author and date for his source *New London Dispensatory*, but may be referring to that of the same name by the prolific but somewhat disreputable William Salmon. Salmon's *The Pharmacopoeia Londininensis. Or the New London Dispensatory* in various editions is the only record for *The New London Dispensatory* in this period either at the British Library or the Wellcome Institute for the History of Medicine.

[50] Samuel Dale, *Pharmacologia, seu manductio ad materiam medicam, in qua Medicamenta Officinalia Simplicia. Hoc Est Mineralia, Vegetablila, Animalia earumque partes in medicina Officinis usitata, in methodum naturalem digesta succincte & accurate describunter*, London, 1693, 1705, rev. 1710, supplemental 1718, 3rd edn 1737, 4th edn 1739.

to the printers.[51] In any event, Blair was doing little more than rearranging the existing knowledge of plants and adding his own additions of novelties, curiosities and miscellaneous observations. William Sherard, an expert botanist, also examined Blair's *Pharmaco-Botanologia* in manuscript, and summarily dismissed it as: 'not to my purpose'. Sherard also questioned Blair's botanical competency to properly identify plants, and criticised: 'he is very large and particular about the sea-wormwood, of which he makes about twenty sorts; which, however distinguishable when growing, are not distinct by the specimens he has sent of them'.[52] What is more serious, Blair's indeterminate 'mixed method' for arranging plants constituted a complete reversal of the established procedures for botany within the community, which were concerned above all with the discovery and identification of plants and, in particular, previously undescribed plants. Further, Blair's rejection of the culturally dominant Raian classification combined what appeared to be an uncritical and arbitrary intermingling of contrary of taxonomic principles, and seemed to threaten to return botany to a state of undisciplined disorder.

Meanwhile, one of the new recruits to botany during the 1720s was Samuel Dale's nephew, Thomas, who had studied medicine under Boerhaave in the Netherlands. In 1723, Thomas dedicated his medical thesis, *Dissertatio Medico-Botanica Inauguralis* (Leiden, 1723), to his father and to his uncle, Samuel, as well as to Boerhaave, Sherard, Dillenius and John Martyn.[53] In part, however, the thesis was also a response to what was seen by many to be to Blair's unfounded attack on Ray in the *Botanick Essays*;[54] Blair certainly thought so. Dale himself declared the opinion that there was 'no comparison to be made between the learning of Ray and Morison'; furthermore, he

---

[51] Blair, 'Preface' (1723–28), iv.

[52] *Correspondence of R. Richardson* (1835), letter from William Sherard to Richardson, 13 October 1722, 189.

[53] Thomas Dale, *Dissertatio Medico-Botanica Inauguralis de Pareira Brava et Serapia Off*, Leiden, 1723.

[54] *Ibid.* Blair especially objected to Dale's comments on page 11:

> Quandoquodem Laudis perpetuo Comes est invidia, & quum semper sit impotentis animi, quicquid exaequare non potest, omnibus modis elevare, non desuerunt, qui de summi illius viri Honore & Fama detraxere. Praesertim scriptor quidam proletarius, qui, famae, si quam habet, sui Civis nimium studiosus, manes ejus turpiter & acerebe nimis insectatur & exagitat. In Thessalum suum habuit Galenus, nec Harvaeo desuit Parisanus, & si Rajus ipse in vitam rediret, novi exorirentur Blairei & Morisoni. Atqui haec Res ulterius non urgenda est, sed ni istos detrectatores Calamo perstrinxissem, Principis Herbariorum famae non viderer satis consuluisse. Nunc eo, unde digressi sumus, revertamur. Si hanc Methodum secuti essent Herbarii veteres in suis plantis describendis, nobis sane pulchrius fuisset, in tam densis certe non versaremur tenebris. Illi vero descriptiones adeo breves & confusas nobis tradidere, ut eorum sententiam, semper expiscari nequeamus, sed saepissime egemus Oedipo, qui illam nobis dilucidaret.

believed that Blair was 'highly to blame for treating Ray's memory in such a manner'.[55] In this respect Dale enjoyed the endorsement of many within the discipline. Herman Boerhaave was also of the opinion that the *Botanick Essays* had been 'rather rich' for Morison at Ray's expense.[56] Writing to William Sherard, Boerhaave's comment that he was 'delighted that the British fulfill a pious duty toward the memory of the good *Rajus* [Ray] who has deserved so well of the public: – for I count it as in the interest of all the Good that virtue be praised after death, and that posterity be commemorated'[57] likely refers to Sherard's own project to sustain the Raian heritage. With the assistance of Samuel Dale, Richard Richardson and the German physician John Jacob Dillenius (FRS 1724 and foreign secretary 1728–47), Sherard was preparing a new edition of Ray's *Synopsis* of British plants. On 25 June 1724, the third edition of *Joannis Raius Synopsis Methodica Britannicarum*, citing Dillenius as editor, received the imprimatur of Isaac Newton, president of the Royal Society. John Martyn also believed that Dale had been urged to write the thesis 'by several others that would not be seen to do it themselves, however in time the truth may be known'.[58] Presumably, Martyn was in a position to know the truth of this. Dale, of course, had dedicated his thesis to Martyn, among others, but Martyn identified Dale as a friend and they were associated as members of a close-knit botanical society in London.

Interpreting Blair's reaction to Dale, however, is complicated by the fact that Blair considered Martyn to be both his protégé and intimate friend; indeed they were frequent correspondents until Blair's death. Blair trusted Martyn to edit his publications and to negotiate with printers on his behalf. For his part, Martyn also attempted to mediate the disagreement between Blair and Dale.[59] Thus, Martyn's own role in the affair is somewhat ambiguous. Nevertheless, Dale's criticism occasioned an extensive and sometimes incoherent correspondence between Blair and Martyn. In July 1724, Blair's fury is evident in their first exchange on the matter:

> Had your friend Dr. Dale only contented himself with discoursing a few theses among his friends in the which I am made the Butt of his malice since they were not otherwise

---

[55] Rawl. MS 323, letter from Martyn to Blair, London, 10 October 1724.

[56] G.A. Lindebom, ed., *Boerhaave's Correspondence*, part 1, Leiden, 1962, letter from Herman Boerhaave to William Sherard, Leiden, 13 April 1720, 88; 'J'ay lu l'ouvrage, qui contient des observations nettes & singulieres, mais n'est il pas un petite pézelé pour Morison, & par consequence un peu moins favorisant au Bon Rajus?'

[57] *Ibid.*, letter dated 5 July 1720, 91.

[58] Rawl. MS 323, letter from Martyn to Blair, Chelsea, 18 August 1724.

[59] Rawl. MS 323, letter from Martyn to Blair, London, 10 October 1724; and an undated letter from Martyn to Blair. An extract from the latter reads: 'I hope you will not think it below your notice I assure you I write him a severe reprimand for abusing a friend he knew I valued in any piece dedicated to me. I shall tell him more of my mind when he returns to England'.

> to be exposed to the view of this work I could have born it. But when not only himself but others ... have done their utmost in the most utrageous [sic] manner to oppose me for standing up in defense of Dr. Morison a Scotsman against the most calumnious and opprobrious reproaches of Mr. Ray, an Englishman who, in a most uncharacteristic manner unworthy of so great a man did continue to load Dr. Morison with personal failings and imperfections for above 20 years after his death and gave occasion to all other forcigners to do the like ... Therefore it is that I must be branded with all the reproaches a youthful calumnious tongue can invent in so publick a place as London among his own country men and mine (as I have been very well assured from thence) thereby to blast and wound my fame so ... and to serve his speech to the utmost height he must search after the most uncommon Ingenious expressions and in the most solemn and publick manner thereby to ensure me the ridicule and mockery of the world.[60]

A consequence of Dale's defence of Ray was that Blair, quite naturally, became concerned about the damage to his own personal credit and reputation. In this and subsequent correspondence, he addressed himself to rescuing Morison's reputation (and so also his own) by repeated assaults on Ray. He also interpreted his remit as one to defend Scots nationality more generally. In fact, Blair's voluminous correspondence on the subject indicates that he had understood Dale's criticisms constituted more general 'national quarrels' with the Scots in addition to the factions and disputes that may have existed within the botanical community. Thus Blair could complain about 'the clamor of English and forreigner [sic] against me. Those must only be young people who know nothing to the contrary or they who are so besotted to Mr. Ray's memory as not to listen even to the naked truth'.[61] Blair, however, may have been justified in his nationalistic concerns. While the Royal Society deployed a rhetoric proclaiming universal values of truth and progress, at the same time it institutionalised a 'jaundiced xenophobia' that coloured its attitudes and acceptance of the non-English.[62] More generally in England, an environment of pernicious national differences also existed between the English and the Scots; it was not uncommon to find endorsement for remarks such as 'the Scotch [sic] and English has always been like Doggs & and Catts'.[63] Especially after the uprising in 1715 to put 'the Pretender' James III on the British throne, English opinion became convinced that Jacobitism posed a very

---

[60] British Museum of Natural History, Banksian MS 60, letter from Blair to Martyn, July 1724, ff. 167, hereafter Banks MS.

[61] *Ibid.*, letter Blair to Martyn, Boston, summer 1726, ff. 281.

[62] Robert Iliffe, 'Foreign Bodies: Travel, Empire and the Early Royal Society of London. Part II. The Land of Experimental Knowledge', *Canadian Journal of History* **34**, 1999, 23–50.

[63] 'A Comment by a Government Agent in Paris', letter from John Semple to Horatio Walpole, Paris, 20 September 1725, quoted from Gregg Edward, 'The Politics of Paranoia', in Eveline Cruckshanks and Jeremy Black, eds, *The Jacobite Challenge*, Edinburgh, 1988, 42–56.

real threat to national stability and the issue features as one of the major concerns in contemporary politics.[64] It may be significant, then, that Blair had been captured in the Jacobite uprising at Preston in 1715, transported along with other notable prisoners to London, imprisoned at Newgate, and sentenced to death for treason.[65] In his own defence, Blair claimed to have been unwillingly forced to assist the rebels, and through the intercession of both the Royal Society and the Royal College of Physicians that his sentence was stayed.[66] However, his education at the high Anglican breeding grounds at St Andrew's University in Scotland has all the hallmarks of a Jacobite, and the active participation of two of his brothers in the uprising also suggests a strong sympathy with the cause. Thus, Blair's Scottish and Jacobite roots may well have been a factor in his reception in London, where his attempts to establish a medical practice were unsuccessful and his efforts to be initiated into the natural history collective largely failed, despite all his hard work to cultivate the patronage of Isaac Newton and especially Hans Sloane.

In any event, by 1725 Blair had few allies within the botanical community, and even John Martyn had become well established in the orthodox network.[67] Indeed, Martyn had begun to offer formal botanical instruction in London, careful to reassure his subscribers that he would proceed 'according to the Method laid down by the late excellent Mr. Ray'.[68] In 1726, Martyn also hoped to teach the course at the University of Cambridge, and published a summary of Ray's method, the *Tabulae Synopticae Plantarum Officinalium. Ad method Raianam Dispositae* (London, 1726) dedicated to Hans Sloane. Although an unofficial, or unlicensed publication, the work was produced by the official printers of the Royal Society. Further, Martyn more adequately reflected the aims and ambitions in contemporary botany, which were to properly identify plants, so that Martyn intended for each plant he demonstrated 'to shew how it differs from others which nearly resemble it'.[69] Praising Ray's classification as the best method for teaching botany, Martyn also criticised the 'inconsistent' mixed method in Blair's

---

64 G.V. Bennett, 'English Jacobitism 1710–1715: Myth and reality', *Transactions of the Royal Historical Society*, 1982, 137–51.

65 Sloane MS 4065, letter from Blair to Petiver, July 8 1716, ff. 253. For Patrick Blair see also Bruce Lenman, 'Physicians and Politics in the Jacobite Era', in Cruckshanks and Black, eds, 1988, 67–79; and A.P. Stevenson, 'Patrick Blair, MD, F.R.S., Anatomist and Botanist', *British Association Dundee 1912 Handbook*, Dundee, 1912, 441–6. For the 1715 uprising more generally, see: Margaret Diane Sanky, *Jacobite Prisoners of the 1715 Rebellion* (PhD diss., Auburn University, 2002); Murray G. Pittock, *Jacobitism*, New York, 1998, 15, 47–8; Leo Gooch, *The Desperate Faction: the Jacobites of North-East England 1688–1745*, Hull, 1995, 91–3.

66 Sloane MS 3322, ff. 145; Sloane MS 4066, ff. 264.

67 Allen (1967), 305–24.

68 John Martyn, *[In April Next Will Begin] A Course of Botany*, London, 1725, 1.

69 *Ibid.*

Pharmaco-Botanologia, in part because an individual would not learn how to identify plants by using it.[70]

Blair's growing isolation in the community was soon manifested in the difficulties he encountered having additional decades of the *Pharmaco-Botanologia* published. By 1727, Blair had achieved five of the originally intended ten decades, and had advanced only as far as the letter 'C' in the alphabet. In responding to Martyn's critique, Blair maintained in Decade IV that 'I could contrive no better means to answer my purpose, than this mix'd Distribution, partly Alphabetical, partly Classical, and both Botanical and Medicinal'.[71] He also continued to promote his original intentions for resolving the differences between the many botanical classification schemes. Since he was 'not being obliged to recommend one Method as more excellent, or decry the other as not so good; but to join them all together, and so to reconcile them, as to shew their Excellence in general, and the great Advantages that attend the teaching by Method, demonstrating whereby that there is no such Discrepancy among them as at first View may appear, or the more ignorant may imagine'.[72] Blair would also claim that Ray's method, as promoted by Martyn, in conjunction with his own combination method, jointly operated to 'lay a solid foundation for the knowledge of botany and *materia medica*'.[73] Intent also on advancing his earlier opinions on the circulation of sap in plants and plant sexuality, Blair returned to these topics in his Decade IV.

The *Pharmaco-Botanologia* was a publishing disappointment and at least part of its failure lay in Blair's unorthodox scientific opinions. Early in 1727, Blair wrote to Hans Sloane requesting his influence to 'prevail with any of the Booksellers' to undertake the project, and again offering to dedicate the work to Sloane. Indeed, to ensure that Sloane was not inconvenienced in the matter, Blair composed the text of a note of endorsement for Sloane to sign and send to the printers.[74] He further suggested that there was no need for Sloane to waste time by actually *reading* the manuscript, 'since it would be too much trouble to you to read the whole'.[75] Besides, Blair claimed, John Martyn had already read the work and approved of it, and, in fact, Martyn was regularly editing most of Blair's works by this time. We are not surprised to find that Sloane preferred to waste his time by reading the manuscript and, since subsequent decades appeared without dedications, it is safe to assume that Sloane was reluctant

---

70 John Martyn, *Tabulae Synopticae Plantarum Officinalium. Ad method Raianam Dispositae*, London, 1726.

71 Blair, 'Preface', Decade IV (1723–28), iv.

72 *Ibid.*, v.

73 *Ibid.*

74 Sloane MS 4048, letter from Blair to Sloane, London, 25 March 1727, ff. 134. Blair's note for Sloane to sign read: 'I desire Dr. Blair's Discourses on Natural History be forthwith printed and published. H.S'.

75 *Ibid.*

to be associated with Blair's opinions. Blair also seemed to finally recognise that the success of his venture rested on the approval of the natural history community; in Decade VI he professed his willingness to 'submit the performance to the censure of the Impartial, Judicious and Learned'.[76] He made modest overtures to the community and began to modify the contents of the work to reflect more of the norms of the discipline, to credit assistance from other botanists, to include illustrations, and to extend discussion on economically important plants. Blair also seemed to acknowledge that the promotion of an unorthodox classification method had not been an acceptable alternative, but felt obligated to continue with his original intentions. Thus Blair wrote, 'I could by no Means depart from my Method ... [and] I should have been injurious to the Publick and wanting to my self, if I had not perform'd what was to be expected'.[77] The eleventh-hour courtship of community approval was unsuccessful, however, and by late that year Blair began seeking private subscriptions to finance the project. He wrote one final letter to Hans Sloane:

> I have neither interest, money nor friends to do anything for me saving your worthy self. I am at present reduced to great weakness that I am scare able to stand alone and as I am very sensable [sic] of your great kindness so far be it from me to desire you to take any such trouble on Yourself but may I beg you please that among your numerous acquaintances you'll please to employ such fit persons as may influence others to encourage the design. It's probable many of the Royal Society and College of Physicians and several of the curious persons in town who otherwise would lye dormant may be brought over by such means when they understand its at your desire ... in a word all is now at stake if these Subscriptions go successfully on I shall thereby be rid of all my present difficulties and be capable of serving those who have been my good benefactors in this matter. If it fails I am utterly ruined.[78]

The last instalment of the *Pharmaco-Botanologia*, Decade VII, appeared in 1728, still enmeshed in the discussion of plants whose names began with letters in the first third of the alphabet. Little more is known of Patrick Blair; indeed, he has become so invisible in the historical records that most accounts also assume his death in 1728. The last we hear of Blair is from a letter written by John Martyn in 1754. 'I very well remember', Martyn wrote, 'that my old friend Dr. Blair was engaged in a controversy with Mr. Bradley, and that not long before his death he sent up some papers on that subject to Mr. Strahan to publish; but on my perusing them I found such marks of delirium, that I persuaded Mr Stahan not to publish them by any means. After his death, some Bookseller talked of publishing his literary correspondence; which

---

[76] Blair, 'Preface', Decade VI (1723–28), iv.
[77] *Ibid.*
[78] Sloane MS, letter from Blair to Sloane, January 27 1727, ff. 93.

some of us used means to obstruct'.[79] Of his own role in the affair, Martyn excused himself: 'At least I can answer for my own letters, which were written when I was a mere boy'.[80]

Blair's case reminds us that the maintenance of scientific culture rests on the shared consent of its local community of specialist practitioners. Indeed, Blair himself recognised too late that the negotiation of differing intellectual opinions or diverging scholarly judgements depended not only upon the commitments of the individual members of the community, but also upon the dedication of those members to perpetuate a particular communal identity and to promote a specific set of disciplinary standards. On the other hand, John Ray's cultural authority as the 'English Aristotle' and founding father of the discipline would continue, and Raian orthodoxy within the natural history community was maintained until the late eighteenth century. Natural history, and especially botany, would be promoted as a legitimate and socially respectable activity, and in particular Raian classification would remain culturally dominant as one of the most important ways of knowing and interpreting nature in England. Thus it was that Ray's natural history project, which served both to define the knowledge domain and delineate the acceptable practices of English natural history in all its varieties and manifestations throughout the century, provided the template for the proper conduct of the discipline within the community.

---

79 Rawl. MS 323, letter from Martyn to Dr Rawlinson, Chelsea, June 16 1754, emphasis in original.

80 *Ibid.*

Chapter 5

Sociability, Utility and Curiosity in the Spalding Gentlemen's Society, 1710–60

Michael Honeybone

This chapter investigates the motivations which led to the development of an extraordinarily successful English provincial learned society, the Spalding Gentlemen's Society (SGS), over a period of 50 years (1710–60). I use the papers of the SGS as a mirror to reflect the changing communication of natural knowledge during the first half of the eighteenth century.[1] The evidence on which this chapter is based is the six surviving minute-books describing SGS meetings (numbering around two thousand in all) held weekly in the provincial town of Spalding in Lincolnshire. The largest number of items discussed at these meetings related to the study of natural history. We can locate in the minutes of societies such as the SGS a system of values which explains the process whereby early eighteenth-century people saw natural knowledge as a significantly useful aspect of their everyday lives. My intention is to offer an account of these values by considering their political, cultural, social and philosophical nature and to examine the way that they were driven by a new early eighteenth-century version of curiosity, whilst still being motivated by a fundamental acceptance of natural theology.

Sociability and Utility

The values in question are those based on natural law, discussed in the work of seventeenth-century philosophers, principally Grotius, Hobbes, Samuel Pufendorf, Locke and Richard Cumberland, and leading to a theory of rights and duties based on the possession of landed property. These values were authoritarian and political in origin, arising from the general distress felt following the civil and religious upsets across Europe in the seventeenth century. It was argued that this society needed to be rooted in the primacy of law, a view derived to some extent from Hobbes, the philosopher whom everyone loved to hate but rushed to read. If one accepts

[1] This chapter is based on research derived from M.J. Honeybone 'The Spalding Gentlemen's Society: the Communication of science in the East Midlands of England 1710–1760' (PhD diss., The Open University, 2001).

E.P. Thompson's thesis, Hobbes' view that self-interest alone controlled social organisations was widely criticised but it none the less led to what might be called a Hobbesian determination to calm down political life, or to the determination to crush upstart radicals.[2] In response to Hobbes, Locke argued that there were more complex laws of nature which could be discovered by sense experience. In the East Midlands, the significant author, Richard Cumberland, bishop of Peterborough, wrote in a similar manner to Samuel Pufendorf who argued that sociability is the 'chief principle in the study of natural law derived from the observation of the nature of things and the desires of men … as it is completely opposed to the Hobbesian doctrine of self-preservation, which Cumberland joins me in attacking'.[3] At the heart of successful groupings of people lay the concept of sociality, defined as the tendency of groups and persons to develop social links and live in communities. Jon Parkin argues that Cumberland's main point was actually somewhat Hobbesian: 'self-preservation necessarily leads to co-operation'.[4] In France the possibility was posed that law could be inherent in social existence: the Academy of Sciences at Dijon set its annual essay competition in 1742 on the question: 'Whether Society can be perfected by natural Laws without the help of Political Institutions'.[5] The new emphasis on sociability also received support from the wide popularity of the philosophy derived from Stoics of the early Roman Empire, which led to calls for civic improvement based on the ancient Roman ideal of *pro bono publico*. It is these two notions of sociality, generally interpreted as sociability, and civic improvement for the public good, that, I suggest, help us to understand how natural knowledge became socially significant: it was perceived as being useful to society. This concept of utility in the early eighteenth century applied also to studying art and aesthetics. Learned activities were expected to be 'useful', in the sense both that they aided study and that they revealed beauty.[6] Together, sociability and utility became the watchwords of the new society.

Curiosity

A route to the study and understanding of the natural knowledge which was perceived to underpin natural laws lay in a renewed awareness of curiosity. As Neil Kenny puts

---

[2] E.P. Thompson, *Whigs and Hunters*, London, 1990.

[3] S. Pufendorf in J. Parkin, *Science, Religion and Politics in Restoration England: Richard Cumberland's De Legibus Naturae*, Royal Historical Society, London, 1999, 209.

[4] Parkin (1999), 100

[5] *Gentleman's Magazine*, November 1742, 607.

[6] Sylvana Tomaselli has pointed out: 'Thanks to the principle of sympathy, according to Hume, we find beauty "in everything that is useful"'. Sylvana Tomaselli, in J.W. Yolton, et al., eds, *The Blackwell Companion to the Enlightenment*, Oxford, 1995, 536.

it: 'The "concept" of "curiosity"... shot to particular prominence from about the mid-seventeenth to the mid-eighteenth century, when it temporarily became central to the construction of both desire and knowledge in various discourses'.[7] The use of 'curiosity' to refer to a quasi-scientific virtue is central to the argument of this chapter. The main new driving force in the communication of the new natural knowledge was curiosity, a word which has today almost lost its specifically seventeenth- and eighteenth-century meaning. The 'curious' (a synonym for 'virtuosi') meant those 'desirous to see and know everything';[8] 'curiosity' was 'a passion or desire of seeing or knowing'.[9] The word has seen several shifts in meaning, which can only be picked up from a detailed study. For many centuries, curiosity was seen as wrong, as it recalled Adam and Eve's transgression. Then, in the sixteenth century, it began to be used to refer to an interest in rarities, as new items of scientific interest were being discovered overseas, and it tends today to be understood as such by historians. One of the few articles on the subject sticks to this meaning.[10] This is inadequate, as the term developed a more general meaning, following Bacon's usage quoted below, being applied to the study of any serious interest. For instance, when plague forced the suspension of classes at Uppsala University around 1710: 'certain professors ... constituted themselves into a "College of the Curious"'.[11] Moreover, in southern Europe, the oldest surviving learned society of the Holy Roman Empire was refounded in 1687 by charter by Emperor Leopold I as *Sacri Romani Imperii Academia Cesareo-Leopoldina Naturae Curiosorum*. Within the early eighteenth-century republic of letters, the 'curious of nature' were those who were, according to Johnson's Dictionary 'inquisitive, attentive, diligent, accurate, careful not to mistake, exact, nice, subtle, artful, rigorous'.[12] This followed Bacon, who was interested in discussing 'the rigour and curiosity in requiring the more severe proofs'.[13]

Those 'curious gentlemen' who formed the Royal Society controlled entry to their ranks; according to Thomas Simpson in 1743: 'the Society (as I am informed) have found it necessary to come to a Resolution not to receive any person as a Member who has not first distinguished himself by something curious'.[14] In effect, the word 'curious' in early eighteenth-century dialogue meant 'precise, careful study' when

---

[7] N. Kenny, *Curiosity in Early Modern Europe: Word Histories*, Wiesbaden, 1998, 13.
[8] N. Bailey, *Dictionarium Britanicum*, London, 1730.
[9] *Ibid.*
[10] K. Whitacre, 'The Culture of Curiosity', in N. Jardine, J.A. Secord and E.C. Spary, eds, *Cultures of Natural History*, Cambridge, 1996, 75–90.
[11] D. Goodman and C. Russell, *The Rise of Scientific Europe*, Sevenoaks, 1991, 313.
[12] S. Johnson, *A Dictionary of the English Language*, London, 1755.
[13] F. Bacon, *The Advancement of Learning*, ed. A. Johnston, Oxford, 1974, 36.
[14] F.M. Clarke, *Thomas Simpson and his Times*, New York, 1929, 92.

used as a noun, and 'worthy of precise, careful investigation' as an adjective. It did not have the full implication of our twenty-first-century term 'science', but it had strong hints of the modern concept 'scientific'. The curious were endlessly satirised, notably in Pope, Gay and Arbuthnot's comedy *Three Hours after Marriage* of 1717. Indeed, by the later part of the eighteenth century, curiosity did indeed begin to acquire the nineteenth-century meaning which Dickens applied to it in *The Old Curiosity Shop* (i.e., appreciation of curious, odd, strange objects, that were interesting but not objects of scientific interest). I prefer to accept Johnson's early eighteenth-century view that curiosity 'is one of the permanent and certain characteristics of a vigorous mind',[15] and that it was a prime motivator in the communication of scientific knowledge during the first half of the eighteenth century.

Thanks to this enthusiasm for curiosity, the period 1700–60 is a wonderful time for the historian interested in the public communication of science. These are the years between the publication of Newton's *Opticks*, immediately accessible in English unlike the Latin *Principia*, and Joseph Priestley's early writings on, and research into, electricity and the nature of gases. It is the time of the great public lecturers, Desaguliers, Demainbray, Benjamin Martin and a host of others who rode around the country with their precious glassware demonstrating physics, mechanics, hydrostatics, light and colours, optics, hydraulics, pneumatics, phonics, astronomy and geography, 'the Whole confirmed by EXPERIMENTS, and illustrated with COPPER-PLATES'.[16] It is a period of exciting new instruments, the sextant, the solar microscope, Harrison's chronometers and Jonathan Sisson's theodolite, to name some discussed in detail at the societies I have studied. New developments in subscription publishing and in engraving were making available better and more intriguing scientific books, pictures and magazines.

The period was above all one of great expectation of advances arising from mechanical philosophy. In the letters and papers of the SGS we find extensive references to the uses of the new philosophy, in particular: pumping, draining, milling, gardening, botany, farming – the application of arts and science to the improvement of nature. The professions were increasingly well-regulated, particularly the law, medicine and the Church, all of which were directly and financially involved in the improvement of the land. These professional men were the new leaders of urban society, replacing out-of-date borough corporations as the driving forces for change. They met in the new provincial learned societies, anxious to learn from the metropolitan Royal Society and Antiquarian Society, but equally concerned to demonstrate pleasure in the study of their own county; as Vladimir Jankovic has

---

[15] S. Johnson, *The Rambler*, no. 103, cited in B. Benedict, *Curiosity: a Cultural History of Modern Enquiry*, London, 2001, 23.

[16] R.V. and P.J. Wallis, *Biobibliography of British Mathematics*, Newcastle, 1986, 164.

expressed it: 'the patriotic politics of chorography was the meta-narrative of Augustan natural history'.[17]

The Spalding Gentlemen's Society

Evidence can be found for the significance of sociability and curiosity to the communication of natural knowledge within English society in the surviving records of local groupings of gentlemen who were concerned to put into practice benevolent ideas of civic improvement by regular meetings.[18] The well-known club movement was recognised by Dr Johnson: 'Boswell (said he) is a very clubable man'.[19] These provincial societies met to discuss learned matters, mainly natural history, antiquities, literature and numismatics. In the East Midlands alone, I have located at least 20 ephemeral early eighteenth-century societies. Generally, they were a phenomenon unique to the period 1710–60, after which they disappeared, to reappear as the literary and philosophical urban societies of the late eighteenth and early nineteenth centuries. The best known was the SGS, founded in 1712 by the barrister Maurice Johnson. In imitation of the well-known London coffee-house clubs, it read weekly publications from London and it catalogued and preserved two existing libraries, those of the parish and the Grammar School. A set of rules provided the procedures of the weekly meetings[20] and every year the SGS's ordinances, or Laws, were revised, re-written and read to the members: 'This Society was instituted for supporting mutual Benevolence, raising and preserving & rendring of general Use a Public Lending Library pursuant to the Statute of the 7th of Queen Ann Chapt 14th. And the Improvement of the Members in All Arts & Sciences'.[21]

The origin of the society's activities was the little-known 1708 Act of Queen Anne, by which endowed parish libraries, by parliamentary statute, had to have a catalogue. During the years 1712–60 there were 381 members of the society, all professional men. About a quarter were Spalding regular members and the remainder were men given

---

[17] V. Jankovic, 'The Place of Nature and the Nature of Place: the chorographic challenge to the history of British provincial science', *History of Science* (2000), **38**, 99.

[18] I would like to record here my thanks to the Council of the SGS for allowing me to use records of the SGS, which are accessible at the SGS, The Museum, Broad St, Spalding, Lincolnshire, PE11 1TB, UK. In addition to the SGS archives, there are two minute-books of the Peterborough Gentlemen's Society for the years 1730–60 in Peterborough City Library. There are two minute-books of William Stukeley's Stamford Society for 1736–37 in the Bodleian Library, Oxford (MS Eng Misc.e.122/123). There are useful papers of the Doncaster Clergymen's Society at Doncaster Archives (Acc.1303 DS33 and P1/5E/1–4).

[19] James Boswell, *Life of Johnson*, ed. R.W. Chapman, Oxford, 1998, 1261, n. 1.

[20] SGS Minute Book 1, folio 40 recto, 1714.

[21] SGS Minute Book 1, folio 217 verso, 1742.

honorary membership in exchange for communicating with the society. Religiously, there were very few dissenters. I can only trace one minister amongst the local Spalding members, the Revd Jonathan Mercer. In addition there were two well known dissenters, Sir Richard Ellys (1682–1742), MP for Boston, widely respected in European learned circles for his classical learning, and John Ward (1679–1758), professor of Rhetoric at Gresham College. By far the largest proportion of members was Anglican clerics, who saw their role as the revelation of the glory of God both through preaching the Word and through the investigation of God's natural world. A large proportion were medical men: Michael Cox, the SGS's operator, was an apothecary and John Green, the second secretary, was a Leiden-trained physician. Table 5.1 at the end of this chapter shows that almost 4 per cent of the members were lawyers and their training encouraged them to look for that order in nature which curiosity might reveal. In this sense, the membership of the SGS was similar to that of the Royal Society; as Richard Sorrenson explained, 'nearly two-thirds of the [Royal Society] membership had to work for a living in one way or another', usually in one of the professions.[22]

Natural Knowledge at the Spalding Gentlemen's Society

The learning discussed by professional gentry in their clubs in the first half of the eighteenth century was wide. The minutes of the SGS reveal that there were regular weekly meetings between 1724 and 1758. The most popular topics were natural history (13.7 per cent), antiquities (11.6 per cent), numismatics (9.8 per cent) and imaginative literature (9.3 per cent). The emphasis on these subjects, as compared with other topics discussed at the SGS, is shown in Table 5.2 at the end of this chapter. The breadth of these topics shows that it was this process of communication of knowledge through curiosity, which was both the *modus operandi* and the chief aim of the society. Table 5.2 also opens up to us those areas of natural knowledge which the gentlemen thought it worthwhile to spend their time on, principally flora and fauna.

It is widely recognised that the study and communication of natural history was at the heart of gentry scientific enquiry in the early eighteenth century.[23] This is

---

[22] Richard Sorrenson, 'George Graham, Visible Technician', *British Journal of the History of Science* (1999), **32**, 203.

[23] Eighteenth-century gentry societies have been thoroughly investigated by Peter Clark in his *British Clubs and Societies 1580–1800*, Oxford, 2000. Two recent collections open up the content of eighteenth-century scientific enquiry, bringing out the extent of the study of natural history: W. Clark, J. Golinski and S. Schaffer, *The Sciences in Enlightened Europe*, Chicago, 1999, and N. Jardine, J.A. Secord and E.C. Spary (1996) (1996). Richard Yeo's book, *Encyclopaedic Visions; Scientific dictionaries and enlightenment culture*, Cambridge, 2001, is also very useful in its discussion of the natural knowledge and natural history contents of the first encyclopaedias.

demonstrated by the huge publishing success of John Ray and the delight in natural theology assiduously propagandised by the Boyle lecturers throughout the century. About half of the natural history topics discussed at the SGS related to plants (or vegetables as they were known in the early eighteenth century) and half to animals. The SGS maintained both a herbarium of dried plants, generally known as a *hortus siccus*, and a garden for plants and medicinal herbs, as part of a significant museum – a concept which in the eighteenth-century meant 'a repository of learned curiosities'.[24] Thus, a 'museum' did not possess then its present-day implication of antiquity. It was a work-place where specimens were collected for active investigation. Such a context allowed plants to be drawn, shown, and grown.

Related to the museum was the physic garden. It was modelled on the work of Philip Miller (1691–1771) on the revived garden of the Apothecaries in Chelsea and on the Physic Garden at Leiden. This setting provided an excellent setting for extensive discussions of the plants and for other activities like the SGS's correspondence with John Green, the Spalding physician who was joint secretary of the SGS. He corresponded with John Harrison, the gardener of Cambridge University, also a member of the SGS, and with J.J. Dillenius, the professor of Botany at Oxford and an associate of Linnaeus. Letters were also frequently received from SGS members in the East and West Indies, giving information on foreign plants and animals and sending specimens for the society's 'museum'. In addition to this, animal, bird and human dissections were frequently discussed and occasionally carried out at the society. Additionally, wax models of the human anatomy were examined and objects prepared in spirits were studied. Correspondents such as John Green, training as a physician in Leiden in 1732, discussed Boerhaave's use of Ruysch's anatomical specimens.[25]

This mention of the Dutch anatomist and surgeon Frederic Ruysch (1638–1731) returns us to the notion of a society interest in natural law and bound together by the newly fashionable curiosity. Ruysch, as the Amsterdam city examiner of midwives, 'had access to monstrous bodies and dissected a remarkable number of them'.[26] The gentlemen at Spalding, a generation later, were likewise fascinated by such curiosities. They recorded many of the famous spectacles – human, animal and mechanical – that were displayed in London and the provinces by travelling showmen; for example, they examined pictures of the 'famous African Hermaphrodite', the 'Bristly Boy' and several very tall and very short people such as a Norfolk 'Lilliputian man' who was 3 foot, 4 inches high. Like the members of the SGS, Ruysch studied specimens such as these for moral purposes. But the SGS members extended these purposes by the addition of a sceptical curiosity regarding the natural world. The minutes of the

---

[24] S. Johnson (1755).

[25] SGS Minute Book 2, folio 62 recto, letter dated 16 March 1732.

[26] William Hagner, 'Enlightened Monsters', in W. Clark, J. Golinski and S. Shaffer, eds (1999), 181.

society record on 7 April 1743 '… what the Showman called Mandrake Male and Female, … the [society's] Operator, Mr Cox, discovered to be nothing else but Roots of the Spanish Anguina wherein the Impostor had inserted dryed Froggs.' [27] What motivated the SGS members here was the new type of curiosity, which could detect inaccuracy as well as locate truth.

Since curiosity was at the heart of the SGS, its orders commanded every member to communicate 'any part of Learning, Knowledge, Arts or Sciences which may to him seem Useful, New, Uncom[m]on, Curious or Entertaining, to be minuted by the Secretaries. Who are to take especial care to carry on all Correspondencies for the sake of the Society'.[28] The learning was to be unusual and entertaining, but it was also to be curious (i.e., precise and useful, or, as we might say today, scientific). It also turned out to be extraordinarily extensive; at the 1,753 weekly meetings between 1724 and 1758, 4,580 topics were discussed. The extent and diversity of these minuted topics set the society a similar task to that being worked on in Sweden in the 1730s by Carl Linnaeus: how to bring systematic order out of the mass of data.

One of the SGS members who would have sympathised with Linnaeus' project was Maurice Johnson. As secretary and then president of the SGS, he tried to bring order to the society's material, but his grand catalogue has a list of members and then hundreds of empty pages. In 1747 he created what became an extensive alphabetical index to the society's six minute-books: 'Farrago Libell[o] Alphabetical Index to the Minutes of the Acts and Observations of Spalding Gentlemens Society'.[29] However, he did not enter items in strict alphabetical order because apparently he could not allot the necessary amount of space as he worked through the minutes. This unsystematic use of the alphabet as a finding device for information across the whole realm of human knowledge proved inadequate. Also he had no model of knowledge which would help him to systematically define individual items within his index. It seems that he did not find it easy to systematise because such a practice was too closely linked to theory. Many in his generation were suspicious of the whole concept of theory, following the Baconian attack on scholastic systems. They believed that the natural order of God's creation and God's providence was enough and questioned whether a new theoretical system of knowledge was either necessary or helpful. This theoretical position was defined in 1730 by the lexicographer Nathan Bailey as 'A system of Philosophy … also a Regular Collection of the Principles and Parts of that Science, into one body, and a treating them dogmatically or in a scholastic way, in contradistinction to the Way of Essay in which the Writer delivers himself more freely, loosely and modestly'.[30]

---

[27] SGS Minute Book 3, folio 142 recto.

[28] SGS Minute Book 1, folio 229 recto.

[29] This volume is today held in the Strong-room of the SGS at Spalding.

[30] The word 'system' as defined by N. Bailey, *Dictionarium Britannicum*, London, 1730.

Curiosity and Systems of Scientific Knowledge

Maurice Johnson's struggle with the concept of systematisation is partly, I would argue, the result of the new values of improvement and utility within a context of sociability. The early eighteenth-century clubbable gentry, driven by curiosity, created museums, libraries, collections of drawings, letters and scientific instruments, and cabinets of curiosities, the existence of which called upon the societies to provide space for their storage and display. This created the problem of how the space was to be ordered. Like Maurice Johnson's index to the SGS minute-books, the approach generally followed was alphabetical (because it was perceived to be non-theoretical) and, indeed, the new encyclopaedias of the time were presented in this form. But as more information flooded in, particularly from overseas members, alphabetical systems seemed too limiting, particularly in the face of new discoveries about the nature, origin and generation of life.

Ordering knowledge was not only limited to the society's collection. The period 1740–50 saw a revived interest in the principle of vitalism. The SGS followed this trend with questions of the generation of life and of the differences between insects, animals and vegetables. Such topics fill the minutes of the SGS at this time. One reason why the SGS spent many meetings on this topic is the high respect held for William Stukeley, an SGS member and a Lincolnshire physician and cleric who lived in London and was a very attentive member of the Royal Society. He sent five volumes of his manuscript memoirs of the Royal Society to be read at the SGS.[31] He gave the society an account of the Royal Society paper on Tremblay's 1740 studies of the hydra – his 'observation on the Insects called a polypus, famed vegetable animal that multiplies without coupling'.[32] Preoccupations with the polyp were being brought to the notice of the European republic of letters in the 1740s. In fact, they were even relevant to the controversy regarding the conception and growth of life in the womb or egg (an argument taking place between the supporters of preformation and those who preferred epigenesis).

The SGS members were fascinated by this vitalism. When discussing this topic, the operator of the society, Michael Cox, a surgeon and apothecary, encouraged another London-based member of the SGS, John Hill, to send his series of published essays, *The Inspector*, to the society.[33] Hill came originally from Peterborough and was a member of both the SGS and the Peterborough Gentlemen's Society. It is in this interest in vitalism where we can see that the SGS was well attuned to debates

[31] These volumes are today held in the Strong-room of the SGS at Spalding.

[32] SGS Minute Book 5, folio 103 verso.

[33] *The Inspector* essays were published in John Hill's weekly, *The Literary Gazette*, London, 1752–54. The SGS were particularly fascinated by *The Inspector*, no. 66, on the Sea Egg or *Echinus Marinus*, read to the society on 9th January 1752.

going on in the capital and abroad. As Emma Spary has explained, preformation was the theory that 'all living things pre-existed as invisible germs, encapsulated one within the other ad infinitum'.[34] The peculiar ability of the polyp to regenerate when cut in half was explained by Haller as arising from this phenomenon. This view was challenged by the believers in epigenesis led by the French naturalist Buffon and the English Roman Catholic priest John Turberville Needham. The latter described their theory (in a paper to the Royal Society) that there existed at conception a range of different particles or 'organic molecules', the heart of the newly revived vitalist theory.[35] In this readable and full Royal Society paper Needham talked of how he had located 'a real productive Force in Nature' as a result of his microscopic studies of a great range of animal and vegetable specimens.

Picking up on these sentiments, Hill's essays used microscopes extensively in his study of spermatozoa and, like his contemporary Needham, he attacked the seventeenth-century preformationist theories of reproduction. Michael Cox and the medical men of the SGS were able to follow Hill's work because they were familiar with the foetuses of both humans and animals. It should therefore come as no surprise to learn that several human foetuses were preserved in the SGS museum. It is also worth emphasising this connection between John Hill's work on spermatozoa and Michael Cox's vitalistic curiosity about the foetus before birth. Both occurred during the most active years of the SGS, in the 1740s and early 1750s. But the increase in knowledge was causing difficulties in the process of relating all this new information to the knowledge structure which was left over from the insistence of late seventeenth-century natural philosophers that their empirical method did not require systematisation.

Based on my comments on the SGS above, it could be argued that it was this new juxtaposition of vitalism with the older seventeenth-century mechanical philosophy which created a need for systematisation to resolve arguments about natural history. This was a major change from the extremely empirical approach of Boyle and Oldenburg in the early years of the Royal Society, well before the establishment of the SGS in 1712. The anti-system position of the early Royal Society had been expressed in a well-known letter by Oldenburg, the first secretary: 'It is our business, in the first place, to scrutinize the whole of Nature and to investigate its activity and powers by means of observations and experiments'.[36] This position was argued for by Newton in his refusal to accept hypothetical arguments and by his application of the extreme voluntarist position. Henry Guerlac explained many years ago that what Newton was doing was 'setting forth … only the *mathematical principles* of natural philosophy. But that philosophy is still to come, the work of many hands, though

[34] Emma Spary, 'Political, Natural and Bodily Economies', in Jardine, et al. (1996), 183.

[35] J.T. Needham, 'Generation, Composition, and Decomposition of Animals and Vegetable Substances', *Philosophical Transactions* **45** (1748), 615–66.

[36] Cited in M. Hunter, *Science and Society in Restoration England*, Cambridge, 1981, 37.

here and there Newton offers hints and suggestions as to what it might contain'.[37] As the alphabetical SGS museum list shows, theorising was unfashionable, as well as increasingly inconvenient, partly because of the intensity of the satirical attacks on natural philosophers: 'the pleasures of essay' (or experiment) prevailed over single-minded system-creation.

It has been fashionable recently to talk of Newtonianism as if it were a single system, but, as Schofield argued in 1970, there was no 'unified and consistent band'[38] of Newtonians. During the first half of the eighteenth century, what we can really see is, as David Haycock pointed out, 'individual Newtonians scattered throughout Britain ... left without constraint to present their separate and idiosyncratic versions of Newton's theory of matter and its action'.[39] The learned societies I have studied were the places par excellence where these idiosyncratic versions were discussed and where everything was grist to their intellectual mills, driven by curiosity.

Oldenburg's famous letter cited above continued with the hope that it was also the business of natural philosophers 'in course of time to hammer out a more solid philosophy and more ample amenities of civilisation'. I submit that, by 1760, this aim, which Bishop Berkeley in his attack on mechanistic philosophy defined as 'speculations to employ our curiosity',[40] was achieved and systematisation again came into play. The communication of new discoveries in the area of natural history was the trigger. Certainly they dominated the discussions on natural philosophy at the SGS. As a result the members had to look for some kind of order. Their desire for order was partly satisfied by the new Linnaean system, which John Hill, an important SGS member, was introducing into England. This practically driven need for order was not foreign to them, as it already existed in their midst in the form of natural theology. This is why they were intrigued by the new vitalistic natural science of Needham. It was based on an underlying notion of a divine creation, which was made accessible by their curiosity and by the microscope. Ordering was therefore fostered by their very curiosity, in particular the solar microscope. Even the new scientific instruments revealed subjects which were beautiful and therefore useful. In the end, the values at the heart of this activity were the new cement of the political order: sociability and utility, fuelled by the new scientific curiosity.

---

[37] H. Guerlac 'Where the Statue Stood: Divergent loyalties to Newton in the eighteenth century', in E.R.Wasserman, ed., *Aspects of the Eighteenth Century*, Baltimore, 1965, 333.

[38] R.E. Schofield, *Mechanism and Materialism: British natural philosophy in an age of reason*, Princeton, 1970, 19.

[39] D.A.B. Haycock, 'Dr William Stukeley 1687–1765: Antiquarianism and Newtonianism in eighteenth-century England' (PhD diss., University of London, 1997), 61–2.

[40] G. Berkeley, *Alciphron*, London, 1732, reprinted in London by Routledge in 1993, 308.

Conclusion

The gentry of the SGS were as empirical as any experimental natural philosopher in their wish to see practical results from their philosophising. Much of their time was spent discussing the technical matters relating to the mathematics of drainage, a crucial issue in their fenland country. They were very much concerned to improve health, as the herb garden and the frequent discussions of physic indicate. They often discussed the practicalities of architecture and topography, as Table 5.2 demonstrates. In these society discussions, they showed their belief that the discoveries that were being made and reported in communications to them were valuable because they were useful. Utility was the surface motivation which gave satisfaction in their sociable interaction. The process of communicating scientific knowledge communication networks, which was developing in the early eighteenth century, took place in a social setting that was being invented *de novo* in the urban trading communities and which needed a sophisticated professional class to order its affairs. It was in this context where sociability became an easily recognised virtue in the polite and mannered society of eighteenth-century Britain. The clerics, lawyers, medical and professional gentlemen, however, needed something else to reconcile them to living close together in the new social settings. It was their shared spirit of curiosity which kept them together in 'curious enquiries' into learned matters. Thus, the style of curiosity which flourished at the time predisposed them above all to investigate natural knowledge.

Table 5.1 Major occupations of the membership of the SGS, 1712–60

Occupation	*Number of SGS members in this occupation*	*Percentage of full membership*
Clerics	90	23.62
Medical	57	14.97
Gentlemen	55	14.43
Merchants and commerce	21	5.51
Lawyers	15	3.93
Painters	12	3.14
Army and Navy	11	2.88
Natural Philosophy and Maths	9	2.36
TOTAL	381	70.84

Table 5.2 Topics communicated to the SGS, 1724–58

Topic	*Number of SGS meetings at which this topic was communicated*	*Percentage of total communicated*
1. Agriculture	12	0.3
2. Antiquities	531	11.6
3. Architecture	127	2.8
4. Astronomy	51	1.1
5. Bibliography	56	1.2
6. Chemistry	15	0.3
7. Education	12	0.3
8. Heraldry	108	2.4
9. Illustrations	403	8.8
10. Imaginative Literature	425	9.3
11. Language	57	1.2
12. Law	40	0.9
13. Mathematics	28	0.6
14. Membership Matters	333	7.2
15. Meteorology	42	0.9
16. Mineralogy	134	2.9
17. Petrifactions	116	2.5
18. Moral/Ethical Philosophy	3	0.0
19. Music	91	2.0
20. Natural History	628	13.7
21. Natural Philosophy	63	1.4
22. Numismatics	450	9.8
23. Physic	165	3.6
24. Religion	96	2.1
25. Seals, Gems, Cameos	256	5.6
26. Technical Matters	215	4.7
27. Topography	105	2.3
28. Trade	8	0.2
TOTAL	4570	99.4

Chapter 6

Set in Stone: Medicine and the Vocabulary of the Earth in Eighteenth-Century Scotland

Matthew D. Eddy[1]

Introduction

Throughout his career, Linnaeus continually paid tribute to the importance of nomenclatural names and descriptions. This is why he included a glossary of terms in almost every edition of his *Systema Naturæ* from 1735 onwards.[2] A large percentage of the systematically minded natural history arrangements during the Enlightenment followed the same practice. The reason so much attention was paid to this subject was because most naturalists knew that the very definitions of a natural object's characters inherently determined how it would be classified. Even though Enlightenment natural philosophers consciously set themselves the task of re-evaluating natural knowledge via the production of dictionaries and encyclopaedias, surprisingly little work has been done to contextualise the actual language they used to describe the terraqueous globe. Even more so than the animal and vegetable kingdoms, eighteenth-century mineralogical vocabulary in the Anglophone world, save for work of the *Oxford English Dictionary*, is a relatively uncharted field of study. Despite the plethora of mineralogical classification systems that existed in 'chymical' writings, lapidaries, herbals, library catalogues and pharmacopoeias, most work on the nascent earth sciences tends to address 'stones' in relation to issues that proved to be of importance

---

[1] Various parts of this chapter were presented at the History of Science 2002 Annual Meeting in Milwaukee, WI, the 2002 'Science and Beliefs' conference at the University of Durham and the History and Philosophy of Science spring 2003 seminar series at the University of Leeds. Further research on this topic was conducted while I was a postdoctoral fellow at the Max Planck Institute for the History of Science (Berlin). In addition to the helpful comments that I received at the above meetings, I would like to thank the following people who gave me helpful advice along the way: David M. Knight, Ursula Klein, Andreas-Holger Maehle, Geoffrey Cantor, Robert Fox, Dane Daniel, John Dettloff, Susan McMahon and Ernst Hamm.

[2] He also included such lists in *Fundamenta Botanica* (1736) and in its later manifestation, *Philosophia Botanica* (1751).

to nineteenth-century chemistry and geology. Such a historiographical disposition is useful for those interested in exploring the foundations of crystallography and historical time, but it offers little for scholars interested in how mineralogy was understood by its eighteenth-century practitioners.

When one begins to delve into the variegated world of early modern mineralogical vocabulary, it becomes quite clear that the transmutation of stone taxonomy into the discipline of nineteenth-century geology was guided by a vocabulary moulded by a background of beliefs which were closely connected to medical perceptions of the human body and theological convictions about matter's relationship with a divine creator. When this vocabulary is explored in closer detail, topics previously considered anomalous to 'scientific' conceptions of the earth, chemico-theology and seminal principles, for example, start to become more relevant. When these ideas are placed in conversation with recent work done on Enlightenment views of probability,[3] chemical epistemology[4] and medical language,[5] the beliefs that shaped the vocabulary of a topic like mineralogy, and hence related topics like chemistry, pharmacology and even cosmology, start to re-emerge as significant epistemological factors for a world where, as Hamm has stated, 'Studies of the Earth's crust were largely empirical and not especially well suited to a model of geometrically demonstratible truth'.[6]

In the past few decades, Laudan, Oldroyd, Rappaport and Emerton have treated several philosophical and experimental issues which are of direct relevance to the vocabulary of mineralogy in the early modern period.[7] These studies suggest that

---

[3] For a re-evaluation of Enlightenment probability in relation to 'geology', see R. Rappaport, *When Geologists Were Historians, 1665–1750*, London, 1997. For mathematical probability, see L. Daston, 'The Doctrine of Chances Without Chance: Determinism, mathematical probability and quantification in the seventeenth century', in M.J. Nye, J.L. Richards and R.H. Stuewar, eds, *The Invention of Physical Science*, London, 1992, 27–50.

[4] W.R. Newman and Lawrence M. Principe, *Alchemy Tried in the Fire: Starkey, Boyle and the fate of Helmontian chemistry*, London, 2002.

[5] Through close study of words in relation to 'rules and norms' consistent with the 'cognitive style' of the eighteenth century, Duden demonstrated that 'keywords like heart, womb or blood connote a dynamic, direct perception'; that is, words considered to be metaphorical by modern scholars were often taken literally in the eighteenth century. See B. Duden, 'Medicine and the History of the Body', in J. Lachmund and G. Stollberg, eds, *The Social Construction of Illness*, trans. J. Mason, Stuttgart, 1992, 39–51, quotations taken from pp. 39 and 40.

[6] E.P. Hamm, 'Of "Histories" by the Hand of Nature Itself', *Annals of Science* (2001), **58**, 311–17.

[7] R. Laudan, *From Mineralogy to Geology: the Foundations of a science, 1650–1830*, London, 1987; D.R. Oldroyd, *Sciences of the Earth: Studies in the history of mineralogy and geology*, Aldershot, 1998; Rappaport (1997); and N.E. Emerton, *The Scientific Interpretation of Form*, London, 1984.

medicine played a central role in the language used to classify minerals. To lay a firmer foundation for future studies on the background 'beliefs' which shaped Enlightenment conceptions of the earth's form and structure, this essay investigates how medicine shaped eighteenth-century mineralogical vocabulary in Scotland. To begin, I will identify three key areas which helped supply the linguistic framework by which the characters and qualities of stones were discussed by physicians in the University of Edinburgh's Medical School: medical Latin, botany and chemistry. Since the conception of a 'stone' was dominated by chemistry, the second section shows how the Medical School's vocabulary (especially that of experimental pharmacology) was applicable to both the human body and mineral composition. Although there are many examples of this linguistic overlap, I focus on how the chemical language of bladder stone composition was transferred into geology via mineralogy. There were many Scottish mineralogists during the mid- and late eighteenth century, but I will concentrate on William Cullen (professor of Chemistry and *materia medica*, 1756–66) and two of his students: Joseph Black (professor of Chemistry, 1766–99) and John Walker (professor of Natural History, 1779–1803). I chose these three professors because they dedicated a notable amount of their lectures to stones and/or mineralogically related topics.

Linguistic Framework

Medical Latin

Almost all of the students who matriculated at the University of Edinburgh's Medical School during the Enlightenment could read Latin. Since lectures were given in Latin until the middle of the century, the scientific names assigned by classical authors were quite authoritative. For those interested in mineralogy, the Latin names of stones and minerals could be learned in several courses, but were specifically taught in the mid-eighteenth-century chemistry and *materia medica* lectures, especially those of Charles Alston and William Cullen. Thus, the mineralogical terminology of Theophrastus, Hippocrates and Pliny were commonly known.[8] These names were re-iterated by most in the Medical School's faculty, including professors Black and Walker, up until the last decades of the century. Even though his mineralogy lectures mentioned

---

[8] For the importance of Pliny's *Historia Naturalis* in history, see E.W. Gudger, 'Pliny's Historia Naturalis. The most popular natural history ever published', *Isis* (1924), **6**, 269–81. The influence of medical Latin in England and in continental Europe is addressed in W. Bracke and H. Deumen, eds, *Medical Latin from the Late Middle Ages to the Eighteenth Century*, Brussels, 2000.

Theophrastus' *De lapidibus*,[9] Walker was more fond of citing Pliny's *historia naturalis*. (This was also the case in his early notebooks.)[10] Pliny discusses minerals in books XXXIII (Metals), XXXIV (Ores), XXXVI (Stones) and XXXVII (Precious Stones), and Walker used many of these names for the genera in his own mineralogical system: 'The number of Genera is exceedingly extensive and it is proper that each should have a name, for this purpose many of the Classical names of Pliny are adopted'.[11] This situation was quite common and Theophrastian and Plinian mineralogical terms were used in most medical communities in Europe. Their continued usage provided a common reference point for an international community of natural philosophers who could read and write Latin.[12] However, since early modern mineralogists had difficulty in identifying the actual composition of Pliny's minerals (as is the case even today), the same Latin names were sometimes applied to completely different stones.[13] Additionally, there was disagreement over how Pliny's terms should be translated into the vernacular. Such hermeneutic issues ensured that local contexts, both intellectual and natural, played a notable role in mineralogical vocabulary.

Despite these translation problems, Latin terms were recycled over and over again in early modern mineralogical works. This was especially true for Plinian

---

[9] For more on how Theophrastus' work fared through the Middle Ages and early modernity, see S.A. Walton, 'Theophrastus on *Lyngurium*: Medieval and early modern lore from the classical lapidary tradition', *Annals of Science* (2001), **58**, 357–79. See also A. Mottana, 'Il libro Sulle pietre di Teofrasto: Prima traduzione italiana con un vocabolario di termini mineralogici', *Atti della Accademia Nazionale dei Lincei* (1997), **8**, 151–234.

[10] Walker cites Pliny's comments on porphyry in the report (*Kings* MS) that he wrote on the Hebrides for George III. See M.M. McKay, ed., *The Rev. Dr. John Walker's Report on the Hebrides of 1764 and 1771*, Edinburgh, 1980, 190.

[11] John Walker, 'Mineralogy Lecture', in H.W. Scott, ed., *Lectures on Geology: Including Hydrology, Mineralogy, and Meteorology with an Introduction to Biology by John Walker*, London, 1966, 229. For more on Pliny's mineralogical names, see J.F. Healy, *Pliny the Elder on Science and Technology*, Oxford, 1999, 115–41, 173–346. See also the 'Index of Minerals' in H. Rackham's introduction to Pliny the Elder, *Historia Naturalis*, ed. and trans. A.H. Rackham, London, 1912, 419–21.

[12] At the University of Edinburgh, students were required to write their medical dissertations in Latin to the end of the eighteenth century. This context is treated in L. Rosner, *Medical Education in the Age of Improvement Edinburgh: Students and apprentices 1760–1826*, Edinburgh, 1991.

[13] D.E. Eichholz briefly treats this problem in the introduction to the Loeb edition of Pliny's *Naturalis Historia*: *Pliny natural history with an English translation in ten volumes, volume X, libri XXXVI–XXXVII*, Cambridge, MA, 1962, ix–xv. A good example of this is Agricola's use of *stannum* in his *De Re Metallica* (1556). In this work, he interpreted Pliny's *stannum* to be the correct term for lead-silver alloys. This definitional difference caused problems for the next two centuries because *stannum* was generally associated with tin, not lead. See G. Agricola, *De Re Metallica: Translated from the first Latin edition of 1556*, trans. H.C. and L.H. Hoover, New York, 1950, 473 and ff. 33.

characters that addressed colour, texture and shape. Walker even cites a few of these terms. For instance, he uses the adjective '*cœcum*' because it was 'a Term used by Pliny, and which, indeed, is very useful in Mineralogy, to denote the lowest degree of transparency'.[14] Not only did Pliny's vocabulary serve as a guide for descriptive Latin adjectives, it also standardised the names given to fossils that were anthropomorphic, zoomorphic or astralmorphic. For example, Walker skipped over the medieval tradition and used Plinian terms to name stones that resembled tongues (*glossopetra*) and stars (*asteria*).[15] Pliny also addressed what Walker and his medical contemporaries saw as chemical characters. The Roman use of heat in metallurgy and saline mixtures in the cleaning and assaying of gems was quite a complex business that resonated with the Medical School's interest in five-principle chemistry.[16] In addition to displaying the typical eighteenth-century anecdotal interest in Pliny's pharmaceutical formulas,[17] Edinburgh's chemical community used his physical and chemical vocabulary to name mineralogical simples and compounds, and to create taxonomical terms. It is for this reason that many of the names that they gave to species, genera, orders and classes had corollaries in Pliny's work.[18]

Botany

A second medical source for mineralogical vocabulary in Edinburgh's Medical School was botany, particularly Linnaeus' *Systema naturae* and *Philosophia botanica*. Even though Linnaeus' botanical system was initially opposed by several members of the medical faculty, particularly Charles Alston, its utility had become accepted by the mid-1760s. By the 1770s, many of the professors, especially Cullen,[19] recognized the

---

[14] David Pollock (transcriber), *An Epitome of Natural History, Vol. IV* (1797), University of Edinburgh Special Collections Library (hereafter EUL), Gen. 706D, f. 26.

[15] For *glossopetra* see Pliny, *Naturalis Historia* (37.164) and for *asteria* see (37.131).

[16] On the interaction of mineralogy and chemistry in classical times, see K.C. Bailey (trans.), *The Elder Pliny's Chapters on Chemical Subjects*, London, 1932; F. Greenaway, 'Chemical Tests in Pliny', in R. French and F. Greenaway, eds, *Science in the Early Roman Empire: Pliny the Elder, his sources and influence*, London, 1986, 147–61.

[17] Even though Pliny gives numerous pharmacological recipes, he is critical about the state of pharmacology in Rome. See Pliny, *Naturalis*, § 34.108.

[18] Pliny's mineralogy is also briefly discussed in R. French, *Ancient Natural History*, London, 1994, 233–40.

[19] As is plainly evinced in the title Cullen's *Nosology, or, A Systematic Arrangement of Diseases: by classes, orders, genera, and species ... and outlines of the systems of Sauvages, Linnaeus, Vogel, Sagar, and Macbride, translated from the Latin of William Cullen*, Edinburgh, 1800. M. Barfoot treats the interaction of pedagogy and method in 'Philosophy and Method in Cullen's Medical Teaching', in A. Doig, J.P.S. Ferguson, I.A. Milne and R. Passmore, eds, *William Cullen and the Eighteenth Century Medical World*, Edinburgh, 1993.

pedagogical advantages of applying Linnaean nomenclature to medical taxonomy. However, the Scots had no problem with accepting Linnaeus' definition and division methodology on the one hand while rejecting several of his classification characters on the other. This meant that there was no one uniform implementation of the Linnaean nomenclature. This had repercussions when Linnaeus' system (which was originally developed for botany) was applied to mineralogy. Throughout Europe, the simplicity of his binomial nomenclature led many naturalists to use his system to classify minerals. A leader in this movement was Linnaeus himself. For terminological simplicity, he had turned to the scientific Latin that he had already developed for botany.[20]

Many of Linnaeus' words were descriptive adjectives that addressed a plant's morphology and/or colour; others addressed essential characters, which, for him, were the parts associated with reproduction. When he began to classify minerals, he decided that the characters by which stones should be classified were, once again, externally observable morphological parts. Thus, the term *rhombus* could not only be applied to the shape of a leaf, but also to the crystalline appearance of a mineral; or, the term *albus* (white) could be used to describe a flower or a type of marble.[21] The dual use of such words was not unique to Linnaeus, but his system had a particularly important impact on botanical and mineralogical vocabulary because it had become a common reference work for naturalists by the late eighteenth century – even for those who disagreed with it. Likewise, most physicians and naturalists in Edinburgh's Medical School had no problem with 'transplanting' botanical terms into mineralogy and numerous examples of this dual terminological usage could be cited. For example, in his early lectures on mineralogical topics, William Cullen used botanical words like foliaceous and fibrous to describe a species of gypsum.[22] Indeed, Walker followed the same practice.[23]

Botanical vocabulary was convenient for those who thought that minerals should be arranged by their externally observable characters; and the wide circulation of Linnaeus' works in Britain and Europe later in the century ensured that these mineralogists

---

[20] The classic source (in English) for Linnaeus' botanical Latin is W.T. Stearn's *Botanical Latin: History, grammar, syntax, terminology and vocabulary*, Newton Abbot, 1973.
[21] *Ibid.*, 311–57.
[22] Cullen used the words foliaceous and fibrous to describe different types of Gypseous Earth. University of Glasgow Special Collections Library (hereafter GUL) Cullen MS 264. He also applied it to muscles during the 1760s: William Cullen, *Lectures on the Materia Medica …*, London, 1773, 14–15. See the *OED* for further eighteenth-century mineralogical uses. Additionally, the word *stamen* (filament, thread) was used by Marcello Malpighi in 1672 and it was subsequently taken up by embryological studies throughout the Enlightenment. H.B. Adelmann, *Marcello Malpighi and the Evolution of Embryology, Vol. I*, Ithaca, 1966.
[23] While travelling in the Hebrides during 1764, Walker noted the similarity between amiantus and petrified wood fibres. Walker (1980), 219. Also see the vocabulary terms in Pollock (1797) EUL 706D, ff. 6–32 and 36–40.

were at least using the same words to describe similar stones. Yet Linnaeus' naturally based, but externally focused, vocabulary presented a few problems for Edinburgh's chemically trained physicians and naturalists, most of whom where heirs to the Becher-Stahl School of chemistry and/or the teachings of Herman Boerhaave.[24] Even though they accepted Linnaeus' definition and division methodology, they based their mineralogical systems on chemical characters. Furthermore, although Linnaeus' basic chemical vocabulary was firmly grounded on the five-principle chemistry that was used in the Medical School (especially in pharmacology),[25] he had only used chemical characters as a last resort.[26] This meant that his system did not provide a vocabulary robust enough to create names for a mineralogical system based primarily on chemical characters. Another standard source was needed to provide names for the chemical characters which could be used to classify minerals. This source ended up coming from the thriving chemistry community in the Medical School.

Chemistry

As Crosland and others have shown, the vocabulary of eighteenth-century chemistry was a complicated affair.[27] Despite this confusion, Edinburgh's professors were united in their belief that there were five basic chemical 'principles': Water, Earths, Salts, Inflammables and Metals.[28] This sort of chemistry was promoted in almost every chemically related course taught in the Medical School: *materia medica*, medical theory, physiology, chemistry, botany and natural history. From the 1750s to the 1790s, the different definitions associated with the key terms of these five principles

---

[24] Many of Edinburgh's mid-eighteenth-century professors had been trained by Boerhaave in Leiden. See E.A. Underwood, *Boerhaave's Men at Leyden and After*, Edinburgh, 1977.

[25] Stearn addresses these chemical terms on pp. 358–63. For the meaning and historical background of pharmaceutical terms, see W.E. Flood, *The Origins of Chemical Names*, London, 1963; J.W. Cooper and A.C. McLaren, *Latin for Pharmaceutical Students*, London, 1950.

[26] He believed this because he held that chemical analysis destroyed a mineral's essential composition. See C. Linné, *A General System of Nature, Through the Three Grand Kingdoms of Animals, Vegetables, and Minerals ...*, trans. W. Turton, London, 1804, 9.

[27] M.P. Crosland, *Historical Studies in the Language of Chemistry*, London, 1962. M. Beretta, *The Enlightenment of Matter: the Definition of chemistry from Agricola to Lavoisier*, Canton, 1993.

[28] To avoid confusion, the names of these chemical principles will remain in upper-case form for the duration of this essay. The methodological and epistemological assumptions that guided these principles in Edinburgh is treated in M.D. Eddy, 'The Doctrine of Salts and Rev John Walker's Analysis of a Scottish Spa, 1749–1761', *Ambix* (2001a), **48**, 137–60 and in A.L. Donovan, *Philosophical Chemistry in the Scottish Enlightenment: the Doctrines and discoveries of William Cullen and Joseph Black*, Edinburgh, 1975.

led to several different types of mineralogical systems in Edinburgh alone (for instance, Charles Alston offered several alternate classifications for minerals in his *materia medica* lectures).[29] Likewise, outside Edinburgh, there were a wide variety of mineralogical systems based on chemistry. These arrangements have been overlooked by historians because they occurred in such diverse sources as museum catalogues, indices in chemistry texts and lists given in *materia medica* lectures.[30] When one looks at the wide variety of these and other chemical mineralogies available to naturalists in Edinburgh, it becomes very clear that there were almost too many sources.

By the middle of the century, the chemically orientated Latin works that emerged as the standard points of comparison for mineralogical characters and vocabulary came from Sweden. After an initial interest in Johann Pott's experiments on Primary Earths,[31] Cullen's work during the 1750s shows that he came to accept the chemically based mineralogical classification of Johan Gottschalk Wallerius, the eminent professor of Chemistry at the University of Uppsala.[32] By the 1760s, he was also entertaining a similar system offered by another Swede, Axel Fredrik Cronstedt. In fact, Walker's personal notes state that it was Cullen who gave him a copy of Cronstedt's *Versuch einer neuen Mineralogie* in 1764.[33] This being the case, from the 1760s onward, Cullen, Black and Walker favourably mention the chemical mineralogy of Wallerius and Cronstedt. The Swedish influence upon Scottish chemical vocabulary was further canonised after the second edition of Torbern Bergman's *Sciagraphia regni mineralis* was made available to the Edinburgh community in 1783.[34] Although Wallerius,

---

29 This was often influenced by how the mineralogical simple was being used in a pharmacological compound. For more on Alston's views on mineralogical simples, see his *Index Medicamentorum Simplicium Triplex*, Edinburgh, 1752, 69–70.

30 See the wide variety of chemical mineralogy sources contained in Walker's library catalogue: Cornelius Elliot, *A Catalogue of the Books in Natural History with a Few Others, which Belonged to the Late Rev. Dr. Walker, Professor of Natural History in the University of Edinburgh*, Edinburgh, 1804. For a set of unpublished lectures that influenced Scottish chemistry, see Charles Alston, *Introduction to Materia Medica* (1736), EUL Dc.8.12.

31 Black also recognised the value of Pott's work in his early lectures: Thomas Cochrane (transcriber), *Notes from Doctor Black's Lectures on Chemistry 1767/8* (ed. Douglas McKie), Wilmslow, 1966, 82.

32 William Cullen, GUL Cullen MS 264. Wallerius' system is explained in Johan Gottschalk Wallerius, *Mineralogié, ou Description Générale des Substances du Regne Mineral. Par Jean Gotschalk Wallerius, Professeur Royal de Chymie, de Métallurgie & Pharmacie dan l'Université d'Upsal, de l'Académie Impériale de Curieux de la Nature*, Paris, 1753.

33 See Axel Fredrik Cronstedt, *Versuch einer Neuen Mineralogie aus dem Schwedischen Übersetzt*, trans. G. Wiedeman, Kopenhagen, 1760. Whether or not this is the copy used by Cullen is unknown.

34 Torbern Bergman, *Sciagraphia regni mineralis secundum principia proxima digesti*, London, 1783. It was also translated during the same year: *Outlines of Mineralogy*, trans. William Withering, Birmingham, 1783.

Cronstedt and Bergman disagreed on several points, their basic vocabulary and systems of arrangement were similar and this allowed their works to become the main source of mineralogical vocabulary in the Medical School.

Chemical Composition

Chemistry and the Earth

So far I have demonstrated that mineralogical vocabulary in Enlightenment Edinburgh came from three areas that were primarily the domain Edinburgh's Medical School: medical Latin, botany and chemistry. Indeed, because of its importance to so many medical subjects (especially chemistry and pharmacology), mineralogy, in some form or another, had been part of the medical curriculum since the school had been founded.[35] In addition, almost every mineralogical classification system developed in Scotland during the eighteenth century was offered by a naturalist who at some point had received a medical education and a good working knowledge of chemistry. This category of mineralogical systematisers included not only Black, Walker, Cullen and their patron Lord Bute,[36] but also those often grouped under the historiographical rubric of geology, especially Robert Jameson,[37] James Hutton[38] and James Hall.[39] Since the vocabulary of mineralogy was so influenced by medicine, this means that

---

[35] R.G.W. Anderson, 'Chymie to Chemistry at Edinburgh', *Royal Society of Chemistry Historical Group Occasional Papers* (2000), **2**, 1–28.

[36] Bute's scientific background and interests are discussed in D.P. Miller, '"My Favourite Studys": Lord Bute as naturalist', in Karl W. Schweizer, ed., *Lord Bute: Essays in Re-Interpretation*, Leicester, 1988, 213–39.

[37] Aside from studying with Walker and Werner, Jameson held a medical doctorate from Edinburgh. He published many mineralogically-related works during the nineteenth century, two of his more well-known being: *System of Mineralogy* ..., Edinburgh, 1804–1808, and *Manual of Mineralogy* ..., Edinburgh, 1821.

[38] Hutton studied medicine for three years in Edinburgh and then went to Holland where he took his medical doctorate in Leiden in 1749. Unsurprisingly, he maintained his own mineralogical collection: Jean Jones, 'The Geological Collection of James Hutton', *Annals of Science* (1984), **41**, 223–44; and he based his conception of mineral formation on chemistry: P.A. Gerstner, 'The Reaction to James Hutton's Use of Heat as a Geological Agent', *Isis* (1971), **62**, 353–62. Even on the last day of his life, he occupied himself by writing down remarks on a 'new mineralogical nomenclature'. J. Playfair, 'Biographical Account of the Late Dr. James Hutton, F.R.S. Edin.', *Transactions of the Royal Society of Edinburgh* (1805), **5**, 39–99, 88.

[39] Sir James Hall (as well as John Playfair) had studied under Walker in 1782. See M.D. Eddy, 'The University of Edinburgh Natural History Class Lists, 1782–1800', *Archives of Natural History* (2003), **30**, 97–117.

these men employed the same words and terms as the physicians who were performing experiments upon stomach acids and who were developing new pharmacological cures. This created a linguistic context in which the vocabulary used to describe minerals and the human body were the same.

This overlap becomes quite significant when one considers that eighteenth-century medical language was often not as metaphorical as modern scholars have assumed.[40] This is a particularly important point because it was the Medical School's chemical mineralogy which laid the conceptual foundation for geology in Edinburgh. This is clearly evinced in John Walker's geology lectures.[41] Indeed, Joseph Black lectured on the chemical aspects of geology throughout his entire career.[42] Such a situation means that it was quite easy for Edinburgh's naturalists to draw analogies between minerals harvested from the Highlands and the chemical experiments being conducted in the Medical School. Traditionally, historians who have looked at chemistry's impact on eighteenth-century Scottish geology have trained their gaze towards experiments that involved high levels of heat. The obvious reason for this being the later success of Playfair's edition of Hutton's *Theory of the Earth*. Yet, a quick browse through the lecture notes taken at the feet of Edinburgh's medical professors demonstrates that the prominent form of analysis was humid, that is, it utilised Salts (acids and alkalis). This was because late Enlightenment medical theory was based upon a form of neo-humouralism which maintained that the fluids of the body needed to be properly balanced.[43] This being the case, Edinburgh's medically focused forms of saline analysis had a direct impact on mineralogical vocabulary that would become of foundational importance to geology. There are many examples that could be used to illustrate this claim, but in what remains of this essay I will present a case study of bladder stones which shows why the geology of Edinburgh's naturalists should be seen through the lens of the chemistry practised by its medical community.

---

[40] See Duden (1992).

[41] Walker (1966). The chemical background of these lectures is treated in M.D. Eddy, 'Geology, Mineralogy and Time in John Walker's University of Edinburgh Natural History Lectures', *History of Science* (2001b), **39**, 95–119.

[42] His lectures during the 1760s referenced many different types of geological formations, and even went so far as to define strata at the beginning of the section on Primary Earths, see especially his comments on earthquakes and the deluge: Cochrane (1966), 55, 144, 165–8. See also his comments about the stratigraphical occurrence of Gypseous Earths, Calcareous Earths and clay on pp. 77–8. Likewise, his *Elements* (1803) is interspersed with many comments on geology. As John Dettloff has pointed out to me, geology and mines were also discussed in French chemistry lectures, particularly those of Macquer and Roulle.

[43] Cullen valued the fluids of the body because he felt that they influenced the 'laxity' or 'rigidity' of the 'solid' parts of the human body. This meant that he believed them to be of central importance to pathology. Cullen (1773), 7–9.

Experimentation and Fieldwork

Bladder stones masqueraded under a variety of names during the Early Modern period. The most common appellations were *calculi* and the 'Stone'; but they were also called 'Earth of Animals', 'Animal Substances' and 'Animal *calculi*'.[44] As Maehle has shown, and as is so clearly evinced in Edinburgh's *Essays and Medical Observations*,[45] bladder stones received a great deal of attention in the Medical School during the entire eighteenth century.[46] Likewise, many of the most prolific chemists in Enlightenment Europe had published at least one tract, essay or letter on the subject. Some, like Wallerius, even included *calculi* in their mineralogical systems.[47] Even though these stones could be surgically removed via a lithotomy, the safer option was to dissolve them via chemical means. This could be done two ways. The first used a syringe to inject a solution up through the urethra and into the bladder. The second utilised oral remedies, either in the form of dietary regulations or via medicines (called lithontriptics) that dissolved and/or dislodged the stones. In order to know which Salts or Earths were most likely to work for any of these remedies, the Medical School performed a barrage of tests *in vitro*. These experiments considered *calculi* to be composed in the same manner and by the same matter as stones found in nature. By employing the same acids and alkalis on mineralogical specimens that they used to test *calculi* (especially Metals and Earths), the vocabulary of medical chemistry was implicitly transferred into mineralogy. The compositional verisimilitude between bladder and mineralogical 'stones' is well evinced by fact that physicians, Cullen's student William Hunter for instance, included *calculi* in their natural history collections.[48] Moreover,

---

44 'Earth of Animals', see fold-out chart in the 'Chemistry' entry in William Smellie, ed., *Encyclopædia Britannica; or Dictionary of the arts and sciences, Vol. II*, Edinburgh, 1771; 'Animal Substances', which Joseph Black seldom got around to discussing completely in his early lectures, Cochrane (1966), 190; 'Animal *calculi*', Stephen Hales, *Statical Essays: Containing hæmastaticks ... also an account of some experiments on stones in the kidneys of the bladder; with an enquiry into the nature of those anomalous concretions*, London, 1732, 190. Additionally, Cullen sometimes discussed bladder stones under the disease category 'nephritics' and treatments involving 'diuretics'. See related entries in the index of *The Works of William Cullen, vols I and II*, Edinburgh, 1827.

45 This was the main publication of Edinburgh's Philosophical Society and the Medical School. See *Essays and Medical Observations, 5th Edition*, Edinburgh, 1771.

46 A.H. Maehle, *Drugs on Trial: Experimental pharmacology and therapeutic innovation in the eighteenth century*, Amsterdam, 1999.

47 The best overview of their placement in his mineralogical classification can be seen in the introductory tables: Wallerius (1753), xxiii, xxxiii–xxxiv.

48 Hunter's *calculi* collection is briefly treated in W.D.I. Rolfe, 'William and John Hunter: Breaking the great chain of being', in W.F. Bynum and Roy Porter, eds, *William Hunter and the Eighteenth-Century Medical World*, Cambridge, 1985.

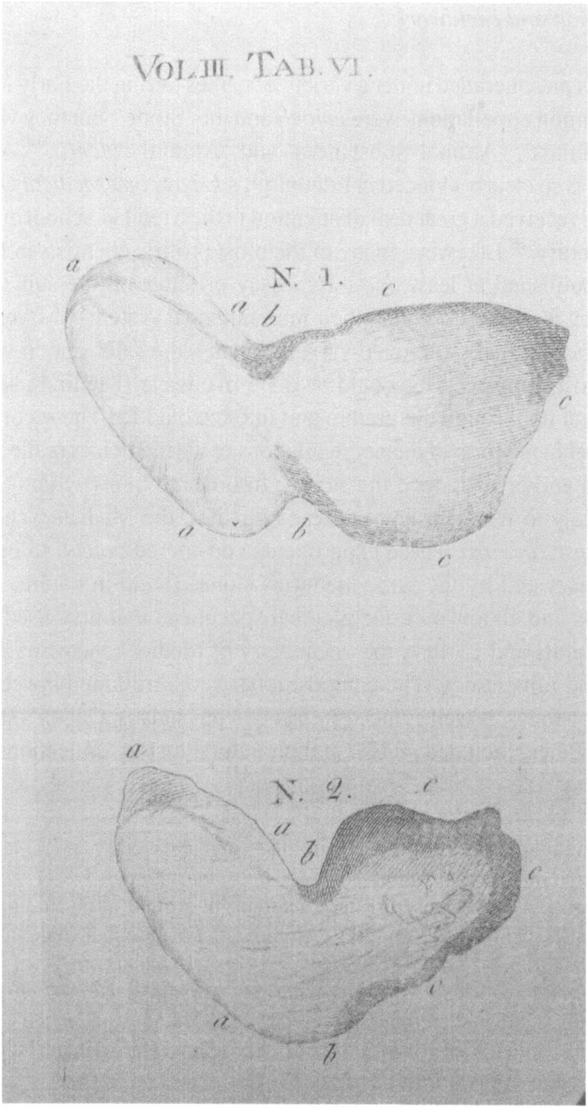

Figure 6.1 A bladder stone as pictured in the 1771 edition of Edinburgh's *Essays and Observations Physical and Literary*, the city's premier scientific journal. It was included to illustrate an article entitled 'The History of Two Cases of Stones Lodged Partly in the Bladder, and Partly in the Urethra' written by 'Dr Livingston', a physician in Aberdeen.

the surgeons of Edinburgh's Royal Infirmary also kept a 'collection of curiosities' that included animal and human concretions. Like *calculi*, these were subjected to chemical experiments and were described with the same vocabulary used for minerals: '[T]here is a ball taken out of the stomach of a horse, which is nearly spherical, and nineteen inches in circumference. Its surface ... being composed of a number of hemispherical knobs ... [and] their outward shell looks like a thin crust of sandy clay'.[49]

To merely dissolve *calculi* and other biological concretions *in vitro* was only the first step. The second objective, in the words of Stephen Hales, was to find 'Menstruums powerful enough to dissolve Metals and Stones, yet so mild as not to hurt, or offend, the tenderest part of the Body'.[50] For the Scots, this came in the form of limewater, an alkaline solution made out of Calcareous Earth.[51] The most common source for this type of Earth was limestone. This created a situation during the last half of the century where many students who had been trained in the Medical School were interested in locating minerals for pharmacological reasons. Cullen led the way with his forays into nature, but he would be followed by Walker, Black, James Henderson and a variety of others. Because of limewater's use in *calculi* experiments, medically trained naturalists were quite keen to identify limestone deposits or mineral wells that contained varying levels of Calcareous Earth.[52] This can be seen in the Highland and Hebrides travel notes taken by Walker and Henderson.[53] (Additionally, Black frequently discussed how to locate minerals in strata in his 1760s lectures). This fostered a context in which observations of Scotland's geological terrain were being made in relation to minerals that were required to prepare saline and terrene solutions which were being used for *calculi* research.[54] As a result, the experiments carried out on the various types of

---

[49] A. Monro, 'Histories of Tophaceous Concretions in the Alimentary Canal', *Essays Physical and Literary* (1756), **2**, 351.

[50] Hales (1732), 190. In addition to being recommended by Black in his lectures, this was a commonly cited source in eighteenth-century Edinburgh.

[51] The production of limewater engendered a heated debate between Charles Alston and Whytt and they both published vociferously on the topic during the 1750s and 1760s. This spurred related essays by other authors in the *Medical Essays and Reviews.* Black mentioned 'The disputes that arrose bewixt Dr Whytt & Alston' on several occasions in his 1760s lectures. Cochran (1966), 65.

[52] See Black's 1760s comments on mineral water, Cochran (1966), 165–70. Using mineral water to dissolve bladder stones went back to ancient times, but began to receive concerted 'chemical' attention at the end of the Middle Ages. See L. Daston and K. Park, *Wonders and the Orders of Nature 1150–1750*, New York, 1998, 135–59. For the Early Modern period see Roy Porter, ed., *The Medical History of Waters and Spas*, London, 1990.

[53] Walker (1980). James Robertson, *A Naturalist in the Highlands*: *James Robertson, his life and travels in Scotland 1767–1771* (ed. D.M. Henderson), Edinburgh, 1994.

[54] Since I am concentrating on *calculi*, I will refrain from commenting on the chemical efficacy of pharmacological observations made on mineral wells (Salts) and mines (Metals).

Calcareous minerals collected in the field would go on to influence how the Scots viewed the chemical composition of geological formations. A good example of this knowledge transfer is evinced in Black's chemistry lectures where he uses the results from *calculi* experimentation to explain the composition of limestone strata.[55]

Stone Formation

The medically engendered conceptions of mineral composition discussed above also led to background beliefs about stone formation, particularly in relation to 'gluten', 'incrustations' and 'concretions'. One of the best examples of where the chemical vocabulary associated with *calculi* experimentation had a direct impact upon Scottish conceptions of geological composition occurs in Black's 1767 chemistry lectures on absorbent Earths. When discussing the composition of limestone, he states: 'The Calcareous Earth is most generally the Matter whereby bodies[,] Vegetable or Animal Substances[,] that have been exposed to water become Stones. w^{c} shows that it can be keep't in water of a Solid or fluid form & it's by this the diff^{t}. waters petrify bodies'.[56] He then states that the substance which concretes these stones together is 'animal Glutinous Matter'. Here the word 'Glutinous' is of particular note. It came from the Latin term *gluten* (the French form being *glutineux*)[57] and was commonly used in medical circles to describe the sticky material that was often produced by putrefaction experiments conducted on both animal and plant remains.[58] That Black transferred it to limestone is not surprising, since he clearly indicates that he believed Calcareous Earth came from compressed shells. What is notable is that, two decades later, Walker used the terms 'gluten' and 'congluten' to describe the cemented matter that held together the rocks which were found in *primary* strata (e.g. the oldest rocks of the earth). These rocks

---

[55] Black specifically states that his original experiments 'into nature of *magnesia*' were conducted because he wanted to find 'a new sort of Lime-water, which might possibly be a more powerful solvent of the stone than commonly used'. Joseph Black, *Experiments upon Magnesia Alba, Quick-lime, and Other Alcaline Substances ...*, Edinburgh, 1777, 6–7. Limewater's connection to *calculi* is not as clearly stated in his *Lectures on the Elements of Chemistry*, Edinburgh: 1803 (which might be a result of editorial omission). In the sections on this subject, however, Black clearly addresses 'calcareous earth in its natural state' (e.g. limestone) and refers the reader to sections of the *Edinburgh Pharmacopoeia* which address the preparation of limewater for the treatment of *calculi*. Black (1803), 36–8.

[56] Cochrane (1966), 57. 'Stones are bodies insipid, hard, not ductile or malleable, nor soluble in water.' Johnson, *A Dictionary of the English Language: First and fourth editions* (ed. A. McDermott), Cambridge, 1996.

[57] The first and fourth editions of Johnson (1996) state: 'GLUTINOUS. *adj*. [*glutinex*, French, from *gluten*, Latin.] Gluey; viscous; tenacious'.

[58] Hence, 'glue', which served as a cognate of 'glutinous' in the Early Modern period. Cullen also used the terms 'gummy' and 'resinous'. Cullen (1773), 2.

contained no biologically engendered glutinous material because they were devoid of extraneous fossil remains.[59] This means that he had transferred a chemical term from its usage in Edinburgh's medical world to the newly emerging field of geology. This is a particularly significant point because Walker was the first professor in the Medical School to designate 'geology' to be a separate topic within his syllabus.

The fact that Edinburgh's chemists thought *calculi* and several different types of minerals were made out of the same kinds of concreted matter was not unique; in reality, this had been the case in Britain since at least the end of the seventeenth century.[60] It was generally believed that Salts (both from mineral wells and ocean water)[61] and even corals[62] could produce indurated stones that were just as chemically and mechanically resilient as rocks dug out of the earth or compounds synthetically produced by heat analysis. Moreover, the formation and composition of *calculi* and indurated stones took place over a relatively short period of time. Such sentiments about formation of stony substances might appear of limited significance to the nascent earth sciences until one considers that the professors of the Medical School were transferring chemical language into geology via mineralogy. Walker's use of gluten is the best example, but other words like 'concretions', 'petrefactions', 'congellations' and 'incrustations' were also imported into various lectures that both he and Black gave on the chemical composition of strata.[63] Thus, the vocabulary and associated

---

[59] This belief about primary strata was generally held by the chemists in the Medical School. See Black (1803), 167; Walker (1966), 175.

[60] For instance, Hales drew no distinction between 'tartarine' incrustations formed in the human body and in mineral wells. Hales (1732), 236.

[61] This point is well illustrated in a letter written from Edward King to Sir John Pringle (a Scott commonly cited in the lectures of both Cullen and Walker) in the 1779 edition of the *Philosophical Transactions*. This *communiqué* addressed the 'petrefactions' and 'incrustations' that had formed around the rope of a ship that had been wrecked off the cost of Scotland in 1745. King asserted: 'The substance of the rope is very little altered; but the sand is so concreted round it, as to be hard as a bit of rock … just in the same manner as impressions of extraneous fossil bodies are often found in various kinds of strata'. Edward King, 'Account of a Petrefaction Found on the Coast of East Lothian', *Philosophical Transactions of the Royal Society* (1779), **69**, 36.

[62] In reference to corals found on a shipwreck described by Sir Hans Sloan, Black stated: 'these Shelly Concretious matters float in great quality in the Sea'. Cochrane (1966), 58. Interestingly, based on both artificial and natural criteria, corals were classified as stones in seventeenth-century pharmacological systems and lapidaries. See Robert Lovell, *ΠΑΝΖΩΟΡΥΤΟΛΟΓΙΑ. Sive Panzoologicomineralogia. Or a compleat history of animals and minerals …vol. I*, Oxford, 1661, 71; Thomas Nicols, *A Lapidary or History of Precious Stones …*, Cambridge, 1652, 160–65.

[63] Especially the concept of 'concretion'. Pollock (transcriber), *An Epitome of Natural History Vol. VI* (1797), Gen 708D, ff. 59–63. Additionally, Black described Jasper as a 'concretion' during the 1760s. Cochran (1966), 80.

Figure 6.2 Professor Joseph Black (1799) as pictured in his posthumous *Lectures on the Elements of Chemistry*, Edinburgh, 1803.

concepts of chemistry implicitly affected the Medical School's perception of geology and suggested that a great many of the Earth's minerals could be formed within the terrestrial time spans suggested by the Bible and other classical texts[64] – and, in this sense, scientific experimentation and theological beliefs could exist side by side within the same epistemological framework.[65] Such a context also explains the rocky reception to Hutton's 'Theory of the Earth' paper when it was read before the Royal Society of Edinburgh in 1785. Even though it was based on chemical concepts, it simply did not fit within the experimental culture's perception of the earth's composition (or heat for that matter).[66] It therefore is not co-incidental that his ideas became more probable in the early nineteenth century, a time when the Medical School's influence over the language of mineralogy and geology was becoming weaker.

---

[64] The authoritative role played by classical texts (a category that she takes to include the Bible) in the nascent earth sciences is treated in Rappaport (1997).

[65] For more on Walker's conception of the earth's age, see Eddy (2001b).

[66] P.A. Gerstner, 'James Hutton's Theory of the Earth and his Theory of Matter', *Isis* (1968), **59**, 26–31.

Conclusion

This essay has addressed the medical vocabulary of mineralogy in Enlightenment Edinburgh. The first half concentrated on three larger linguistic considerations – medical Latin, botany and chemistry – while the latter part used bladder-stone research to show how the vocabulary of chemistry was transferred from medicine to geology via mineralogy. As I have indicated in several of the footnotes along the way, this essay only treats a fraction of the eighteenth-century mineralogical vocabulary that was affected by a medically orientated epistemology. For instance, for bladder stones alone, several other topics could have been pursued – especially how cross-sectional representations of their contents closely resemble early stratigraphical drawings, both of boulder stones and mountains. Furthermore, *calculi* were sometimes called petrefactions, a term also used to identify extraneous fossils. Aside from bladder stones, the vocabulary of botanical characters was also employed by the Scots to delineate the external features of stones and this allowed words (and their associated definitions) from anatomy and physiology to be used to describe the physical structure of geological formations and phenomena (especially volcanoes and earthquakes). Furthermore, medical vocabulary not only interacted with the observations by physicians who toured the countryside, it was also tested against what was being dug out of the many mines found throughout Scotland. The latter point is particularly important because Scotland did not have any mining academies and this meant that the Medical School's professors played an important role in assaying ores.

In the 1966 foreword to his first book, David M. Knight wrote: 'If we are to assess scientists of the past, we must judge their views ... by their consistency and their power to explain the phenomena then known and felt to be puzzling'.[67] In many ways, my comments about the medical vocabulary of mineralogy in this essay have addressed the 'consistency' of language in a historical context where words migrated between studies which treated both human and earthy bodies. This has allowed me to show that the mineralogically minded Scots addressed in this study were very interested in the form and the structure of the earth, but not in the way which has been portrayed by most histories of the earth sciences. When these men pondered the terraqueous globe they were puzzled, in an experimental sense at least, by questions of systematisation and medical utility – and this no doubt explains why the bulk of laboratory work in late eighteenth-century Edinburgh revolved around issues that were relevant to experimental pharmacology. As several other essays in this collection also mention, asking questions about intellectual milieus in such a manner brings important, and often

---

[67] D.M. Knight, *Atoms and Elements: a Study of theories of matter in England in the nineteenth century*, London, 1967, 1.

intangible, intentions to the foreground and takes us one step closer to understanding the beliefs that motivated early modern natural philosophers to be so inquisitive about the natural world.

Chapter 7

Scientific Servicemen in the Royal Navy and the Professionalisation of Science, 1816–55

Randolph Cock

In 1974, in a paper entitled 'Science and Professionalism in England, 1770–1830', David Knight posed the questions: What was the nature of science in that period? Why did it receive financial support from government? And what opportunities were there to pursue a *career* in science?[1] He pointed out that the armed forces, and especially the Royal Navy, played a significant part in British science, and that naval science was generally 'Big Science', requiring a lot of money and the teamwork of cooperating specialists. The Navy provided significant career opportunities in science, and Knight identified the beginnings of the professionalisation of science in the civilian naturalists and astronomers who accompanied the Navy's Arctic expeditions during the 1820s, along with what he saw as the opposite development, the training of naval officers to carry out scientific research.[2] It may be illuminating to look again briefly at those issues, in relation to the slightly later but overlapping period roughly between the Napoleonic and the Crimean Wars.[3]

---

[1] David Marcus Knight, 'Science and Professionalism in England, 1770–1830', *Proceedings of XIV International Congress of the History of Science, 1974*, 2 vols, Tokyo, 1975, **1**, 53–67.

[2] It may be interesting to note in this connection that whilst the Royal Navy continued to employ a mix of civilian and naval personnel in their scientific enterprises, the French Navy came to believe that civilians undermined naval discipline and decided in the 1820s to banish them from their ships entirely, all scientific roles being taken over by naval officers, see Hélène Blais, 'Les officiers de Marine français et la géographie du Pacifique dans la première moité du XIXe siècle: entre science et empire', in Pieter van der Merwe, ed., *Science and the French and British Navies, 1700–1850: the 8th Anglo-French Conference of Naval Historians*, National Maritime Museum, Greenwich, forthcoming.

[3] In this paper I employ two themes which enjoyed wide currency among historians of science in the 1970s but which have since had much of the juice squeezed out of them. My object in reviving the desiccated husks of 'Humboldtian science' and that old chestnut, the 'professionalisation of science', is to see whether they do not after all offer a framework within which to explore a rather more neglected but, I believe, at least equally important aspect of early and mid-nineteenth-century British science: the involvement of the Navy in scientific research.

Humboldtian Science and the Division of Labour

The first half of the nineteenth century, especially after the war, was a time when both the nature and the institutions of British science were changing. New scientific societies were being formed in London and the provinces: some, including the Geological Society (founded in 1807) and the Astronomical Society (founded in 1820), were both cause and effect of growing specialisation into distinct disciplines; others, such as the British Association for the Advancement of Science and the provincial Literary and Scientific Societies, were a sign of widening participation in science.[4]

Addressing the question of what characterised science around this time, Susan Cannon introduced the idea of Humboldtian Science.[5] One of the central concerns of Humboldtian Science was the collection of large amounts of data, often using precision instruments, from global or regional surveys. There was a predilection for numerical data, which were often tabulated, or plotted on maps or isoline charts, or given other, usually graphical representations. The collection and plotting of this data were activities for which the training and experience of the naval officer, and especially the naval surveyor, fitted him well, and the expansion of the Royal Navy's Surveying Service under Thomas Hurd (Hydrographer of the Navy 1808–1823), Sir Edward Parry (1823–29), and, particularly under Sir Francis Beaufort from 1829 to 1855, provided a host of opportunities for systematic observations and collections to be made around the world.[6]

Although doubts have been raised by several historians of science about the usefulness of such 'vague generalities' as Humboldtian Science, and even about the validity of Cannon's analysis of early nineteenth-century science,[7] much of the Navy's work in geodesy, geomagnetism, tides, meteorology and ocean currents, carried out in conjunction with hydrographic surveys or voyages of exploration, fits well the Humboldtian paradigm, as do their studies of the large-scale distribution of species of plants and animals, geology, and even, in some ways, archaeology and antiquarian studies. The primacy accorded to the collection of data was apparent in the popular

---

[4] J.N. Hays, 'The London Lecturing Empire, 1800–1850', in I. Inkster and J. Morrell, eds, *Metropolis and Province: Science in British Culture 1780–1850*, London, 1983, 91–119; Iwan Morus, Simon Schaffer and Jim Secord, 'Scientific London', in Celina Fox, ed., *London – World City, 1800–1840*, London, 1992, 129–42; Jack Morrell and Arnold Thackray, *Gentlemen of Science: Early Years of the British Association for the Advancement of Science*, Oxford, 1981.

[5] Susan Faye Cannon, *Science in Culture: the Early Victorian period*, New York and Folkestone, 1978, 73–110.

[6] G.S. Ritchie, *The Admiralty Chart: British Naval Hydrography in the Nineteenth Century*, 2nd edn, Edinburgh, 1995.

[7] James A. Secord, 'The Geological Survey of Great Britain as a Research School, 1839–1855', *History of Science* (1986), **24**, 223–75, 264.

discussions of scientific method by William Whewell and John Herschel (as was also the importance of working hypotheses to guide data collection, which, however, seems sometimes to have been lost sight of in the scramble for 'facts').[8] 'There is no branch of science whatever', wrote Herschel in 1830 in his *Preliminary Discourse*:

> … in which, at least if useful and sensible queries were distinctly proposed, an immense mass of valuable information might not be collected from those who, in their various lines of life, at home or abroad, stationary or in travel, would gladly avail themselves of opportunities of being useful.[9]

Beaufort was a friend of Herschel's, a member of the General Committee of the Society for the Diffusion of Useful Knowledge which published the *Preliminary Discourse*, and as of the previous year, he was in charge of the largest and in many ways best-suited body of men 'at home or abroad, stationary or in travel' who were likely to be useful to Herschel's scheme. The only reason to doubt that he was inspired by Herschel's words to mobilise the Surveying Service for science, is that he had already begun to implement just such a programme.[10]

In his studies of science and manufacturing, Charles Babbage had highlighted the principle of the division of labour, which he believed was equally applicable to many branches of science.[11] Drawing on such ideas from the new machine age, Herschel advanced the opinion that, 'to avail ourselves as far as possible of the advantages which a division of labour may afford for the collection of facts … is an object of great importance'.[12] From the beginning, a relatively well-defined division of labour was implicit in the relationship between the gentlemen of science at home, and their fieldworkers in the Navy: the Navy collected the data, and the scientists analysed it and theorised about it. Thus the development of a class of research assistants was fostered in the Navy. However, in a number of cases, it was not very long before the experience of observing and collecting led the more curious and enthusiastic of this class to begin offering their own, in some cases superior, theories. This development of naval officers and midshipmen into professional scientific specialists happened, for example, in the study of the tides. Initially, in the early 1830s, there was 'clear blue

---

[8] William Whewell, *The History of the Inductive Sciences*, 3 vols, London, 1837, and *The Philosophy of the Inductive Sciences*, 2 vols, London, 1840; John F.W. Herschel, *A Preliminary Discourse on the Study of Natural Philosophy*, London, 1830.

[9] Herschel (1830), 133–4.

[10] Randolph Cock, 'Sir Francis Beaufort and the Co-ordination of British Scientific Activity, 1829–1855' (PhD diss., University of Cambridge, 2003).

[11] Charles Babbage, *Reflections on the Decline of Science in England and Some of its Causes*, London, 1830, and *On the Economy of Machinery and Manufactures*, London, 1832.

[12] Herschel (1830), 133.

water' (in many cases, literally) between, on the one hand, the philosophers William Whewell and John Lubbock, and, on the other, the naval and Coast Guard officers who collected their data. But, in 1836, Captain Robert FitzRoy's observations of tides in the South Atlantic led him to challenge Whewell's theory (which did not fit the data) and to propose his own.[13] Similarly, by the late 1840s, another tidal specialist, Captain Frederick William Beechey, was publishing research in his own right in the *Philosophical Transactions* and the *Admiralty Manual of Scientific Enquiry*, alongside that of Whewell and his colleagues in the scientific community.[14]

Why did the Navy do Science, and Why was it Supported by Government?

Although British science had a proud amateur tradition, the kind of large-scale fieldwork which Humboldtian Science involved was too expensive in most cases for private individuals to fund, and sometimes too complex for them to organise. As James Rennell, a pioneer in the study of ocean currents, and Britain's 'foremost geographer', put the case in 1832:

> Nothing less than a great number of observations of every kind ... can enable the diligent inquirer to make himself master of the whole subject; and this can be the work of the Government only; for individual inquiry can produce little more than unconnected facts.[15]

Only the State could afford to finance such ventures. It was also the case that, in the absence of an established scientific profession, only the State had access to sufficiently large bodies of technically trained manpower.

---

[13] *Journal of the Royal Geographical Society* (1836), **6**, 311–43; Margaret Deacon, *Scientists and the Sea. A study of marine science*, London, 1971 and Aldershot, 1997, 267; David Edgar Cartwright, *Tides. A Scientific History*, Cambridge, 1999, 117.

[14] 'Report of Observations made upon the Tides in the Irish Sea, and upon the great similarity of Tidal Phenomena of the Irish and English Channels, and the importance of extending the Experiments round Land's End and up the English Channel. Embodied in a letter to the Hydrographer. By Captain F.W. Beechey, R.N., F.R.S. Communicated by G.B. Airy, Esq., F.R.S. &c, Astronomer Royal', *Philosophical Transactions of the Royal Society of London* (1848), **138**, 105–16; Captain F.W. Beechey, 'Report on Further Observations upon the Tidal Streams of the North Sea and English Channel, with Remarks upon the Laws by Which Those Streams Appear to be Governed', *Phil. Trans.* (1851), **141**, 703–18; Captain F.W. Beechey, 'Hydrography', in Sir John F.W. Herschel, ed., *A Manual of Scientific Enquiry; Prepared for the use of Officers in Her Majesty's Navy; and Travellers in General*, 2nd edn, London, 1851; facsimile reprint with new introduction by David Knight, Folkestone, 1974.

[15] Quoted in Deacon (1971 and 1997), 235.

By the end of the eighteenth century, the British had developed armies (including the armies of the East India Company) with corps of technical experts including artillerymen and engineers, and the Navy had spawned a Hydrographic Office which was soon to have control of a Surveying Service; together these military surveyors represented a latent resource which could be harnessed to undertake the multi-disciplinary regional surveys which were characteristic of the Humboldtian approach. The Survey of India, the Ordnance Survey, and to a lesser extent the Geological Survey, all depended on military expertise. Matthew Edney has identified an increase in 'map-mindedness' amongst senior army officers from around the 1760s,[16] and it could be said that Humboldtian Science, which by its nature recommended itself to the military, represented the approach of 'map-minded' scientists.

The growth in the employment of the armed forces in scientific work following the precedent set by Cook's *Endeavour* voyage of 1768–71, may be seen as just one aspect of a wider acceptance that it was the State's job to collect information of all kinds. As D. Eastwood has shown, through the agency of a host of parliamentary select committees and royal commissions, legislation in early nineteenth-century Britain was informed by a wealth of social statistics.[17]

There is a school of thought which has attained the status of orthodoxy with many imperial historians and not a few historians of science. Usually unacknowledged but owing something at root to dubious French intellectuals such as Michel Foucault, and tritely equating science with knowledge, knowledge with power, and power with exploitative imperialism, the followers of this school maintain, often implicitly, that science, in whatever period and especially, but not exclusively, science associated with surveys, was 'subservient to the realities of imperial power'[18] and that the scientific servicemen and civilians who conducted surveys around the globe were, at best, the unwitting pawns of malevolent Machiavellian forces.[19]

---

[16] M.H. Edney, 'British Military Education, Mapmaking, and Military "Map-Mindedness" in the Later Enlightenment', *Cartographic Journal* (1994), **31**, 14–20. I am using the word 'army' loosely here, to include the Royal Artillery, Royal Engineers, East India Company forces, etc. With individual exceptions, the British Army itself was not, in this period, a notably scientific organisation.

[17] D. Eastwood, '"Amplifying the Province of the Legislature": the Flow of information and the English State in the early nineteenth century', *Historical Research. The Bulletin of the Institute of Historical Research* (1989), **62**, 276–94. See also the 136 commissions listed in J.M. Collinge, *Office-Holders in Modern Britain,* IX *Royal Commissions of Inquiry 1815–1870, IHR*, London, 1984.

[18] Crosbie Smith, 'William Hopkins and the Shaping of Dynamical Geology, 1830–1860', *British Journal for the History of Science* (1989), **22**, 27–52, 33.

[19] This attitude is apparent in much of the work of Lewis Pyenson (such as 'The Prestige Of Pure Reason: Naval support of science in the Dutch East Indies, 1840–1940', in Paul C. van Royen, Lewis R. Fischer and David M. Williams, eds, *Frutta di Mare. Evolution and Revolution*

To give them their due, it is true enough that in order to effectively exploit the Empire, in staking claim to territory (including dominion over the seas and over the trade borne on those seas which the British came increasingly to see as an extension, indeed, the backbone of their Empire) and in consolidating those claims, empire-builders and colonial administrators were assisted by the act of surveying – both in the narrow sense of cartography and hydrography, and in the wider Humboldtian sense of 'The Survey'. Colonial powers did on occasion apply science to economic and political advantage,[20] and it would be hard to disagree, for example, with Christopher Bayly's assessment of the Survey of India as 'partly an aspect of the assertion of sovereignty', or that 'the survey was also clearly an aid to the practical aim of the collection of territorial revenue'.[21] However, imperial ambitions or considerations are neither a necessary nor a sufficient explanation for the pursuit of science in this period, if in any, and not *all* surveying, and even less science in general, was directed towards the exercise of political, economic or colonial power. The number of foreigners involved in science with the Navy should perhaps raise doubts about why Germans, Danes and others, should have been so keen to participate in the aggrandisement of the British Empire; while widespread international cooperation in science between imperial rivals such as Britain, France and the USA suggests there was more (or less) to the relationship between science and empire than simplistic models suggest. Lobbyists in BAAS were not averse to 'talking up' the national or imperial benefits of particular scientific programmes when campaigning for State support, as they did, for example, in the case of the Magnetic Crusade.[22] But before mid-century, it is doubtful that the acquisition of colonies enjoyed very widespread support in Britain, especially among many of those most attracted to scientific pursuits. The British tradition was one of trade, not large-scale colonisation, and territorial acquisitions made in the wars up

---

in the Maritime World in the Nineteenth and Twentieth Centuries. Proceedings of the Second International Congress of Maritime History, 5–8 June 1996, Amsterdam, 1998, 29–37, and, for example, in Richard Drayton, *Nature's Government. Science, Imperial Britain, and the 'Improvement' of the World*, London, 2000; Smith (1989); and William J. Ashworth, 'John Herschel, George Airy, and the Roaming Eye of the State', *History of Science* (1998), **36**, 151–78. Matthew Edney is typical of many others in making some implicit assumptions along these lines. Also influenced by Foucault, J.B. Harley saw cartography in Early Modern Europe primarily as an instrument of power (J.B. Harley, 'Silences and Secrets: the Hidden agenda of cartography in early Modern Europe', *Imago Mundi* (1988), **40**, 57–76).

[20] Roy MacLeod, 'Introduction', in *Nature and Empire: Science and the Colonial Enterprise*, *Osiris*, 2nd series (2000), **15**, 1–13, 1.

[21] C.A. Bayly, 'Knowing the Country: Empire and information in India', *Modern Asian Studies* (1993), **27**, 3–43, 34.

[22] John Cawood, 'Terrestrial Magnetism and the Development of International Collaboration in the Early Nineteenth Century', *Annals of Science* (1977), **34**, 551–87, and 'The Magnetic Crusade: Science and Politics in Early Victorian Britain', *Isis* (1979), **70**, 493–518.

to and including the Napoleonic, were often bargained away for other advantages, at the subsequent peace conferences. Even to those of the British scientific community enamoured of territorial aggrandisement, the connection between science and empire was not simply that science should serve the ends of imperialists, but that the Empire, and the ambitions of empire-builders, facilitated the collection of data and specimens from around the world. As Knight succinctly observed, 'in nations with an empire to describe, descriptive science will be prominent'.[23]

Whatever the State may or may not have expected from its sponsorship of science, the individuals associated with the Navy who participated in that activity had their own reasons. Many who studied the natural world were concerned simply to reveal the full extent and glory of God's Creation: natural theology (and, indeed, natural curiosity) was as powerful a stimulus to scientific discovery as was colonialism, or trade.[24] The seriousness with which natural theology was taken by many in scientific circles, as well as more widely, was shown by the popularity of William Paley's *Natural Theology*,[25] which went through 20 editions in as many years, and by the commissioning of leading scientists to write the Bridgewater Treatises.[26] One of those, the Oxford geologist William Buckland, for whom, it has been said, 'fieldwork became an explicitly religious experience', argued that 'the real utility of science lay not in economic advantage but in broadening the mind towards knowledge of God'.[27] Also, since 'early nineteenth-century writings on natural theology emphasised the diversity as well as the spatial extension of the living world',[28] the doctrine was an important spur to efforts to catalogue the variety of species as well as their geographical distribution, which were the two concerns of most naturalists who travelled with the Royal Navy.[29]

---

[23] Knight (1975), 56.

[24] On the importance of natural theology in this regard, see Richard R. Yeo, 'The Principle of Plenitude and Natural Theology in Nineteenth-Century Britain', *British Journal for the History of Science* (1986), **19**, 263–82, 266–70; James A. Secord, *Victorian Sensation. The Extraordinary Publication, Reception, and Secret Authorship of* VESTIGES OF THE NATURAL HISTORY OF CREATION, Chicago and London, 2000; Drayton (2000), 234; and Aileen Fyfe, 'The Reception of William Paley's Natural Theology in the University of Cambridge', *British Journal for the History of Science* (1997), **30**, 321–35.

[25] William Paley, *Natural Theology; or, Evidences of the existence and attributes of the Deity ... Collected from the Appearances of Nature*, London, 1802. This work was, for example, included in the Seamen's Libraries issued by the Victualling Office to Royal Navy ships in the 1840s (Public Record Office, Kew (PRO) ADM 114/6).

[26] Yeo (1986), 266; Jon Topham, 'Beyond the "Common Context": the Production and reading of the Bridgewater Treatises', *Isis* (1998), **89**, 233–62.

[27] Roy Porter, 'Gentlemen and Geology: the Emergence of a scientific career, 1660–1920', *Historical Journal* (1978), **21**, 809–36, 821 and n. 55.

[28] Yeo (1986), 268.

[29] Janet Browne, *The Secular Ark. Studies in the History of Biogeography*, London, 1983, 33–44.

Beaufort was clear that surveyors, 'by augmenting the stock of geographical knowledge, … were essentially contributing to the honour of their country',[30] and they were well aware themselves of the benefit to national prestige, as well as the potential economic and humanitarian advantages, of scientific work. They were alive, too, to the possibilities it presented to advance their careers in a time of peace, when opportunities to distinguish themselves in battle were, for the most part, denied them.[31] But it has to be said that many young men chose a career in the Navy simply because they *wanted* a career in the Navy; and many of those who took advantage of the opportunities such a career offered to engage in scientific work, did so because they *wanted* to do science.[32] Whereas Sir Joseph Banks's eighteenth-century contemporaries seldom lost sight of the practical applications of science or its benefits, the early nineteenth century was a time increasingly of 'research for its own sake', guided by 'a research ethic that pursued scientific truth as an end in itself'.[33] John Barrow's appeal to the Navy to pursue 'science for the sake of science'[34] would have struck a chord with many scientists and scientific servicemen. Richard Owen's work in anatomy has been cited as demonstrating that 'science could meet the highest standards … without serving utilitarian or materialist ends'.[35] Similarly, Buckland and his Cambridge counterpart, Adam Sedgwick, 'despite lip-service to utility … would not subordinate their research to narrowly economic, practical problems. They celebrated the intrinsic and ornamental cultural value of scientific knowledge'.[36] And in his *Preliminary Discourse*, Herschel appealed to the well-informed person to go out into the field and observe, not things necessarily of utility or of intrinsic benefit to the Empire, but 'some particular class of facts which most excite his attention' – in other words, whatever took his fancy.[37]

---

[30] PRO ADM 7/847, p. 47.

[31] 'Pax Britannica', of course, was not, however, a period devoid of military adventures, and Britain and her navy were involved in numerous minor conflicts around the world between Waterloo and the Crimea (Bernard Semmel, *Liberalism and Naval Strategy. Ideology, interest, and sea power during the Pax Britannica*, London, 1986, 8; David French, *The British Way in Warfare, 1688–2000*, London, 1990, 120).

[32] As youngsters, Francis Beaufort and Owen Stanley were seized with a passion for science, as, it seems likely, were many others (Alfred Friendly, *Beaufort of the Admiralty. The Life of Sir Francis Beaufort 1774–1857*, London, 1977; Adelaide Lubbock, *Owen Stanley R.N. 1811–1850, Captain of the 'Rattlesnake'*, Melbourne, 1968).

[33] John Gascoigne, *Joseph Banks and the English Enlightenment. Useful knowledge and polite culture*, Cambridge, 1994, 4.

[34] [John Barrow] review of 'A Memoir on the Geography of the North-Eastern part of Asia, and on the Question whether Asia and America are contiguous, or are separated by the Sea; by Captain James Burney FRS, from the Philosophical Transactions of the Royal Society', *Quarterly Review* (May 1818), **18**, no. 36, 431–58, 457.

[35] Morus, Schaffer and Secord (1992), 133.

[36] Porter (1978), 819.

[37] Herschel (1830), 133.

But whatever the reasons why individuals took up an active involvement in science, the Navy was one of the few institutions in which such a pursuit could function as a career, and naval involvement with these investigations owed much to the links which existed between the Navy and the scientific community as represented by the gentlemen of the scientific societies, the universities, and institutions such as the British Museum and the Royal Botanic Gardens at Kew. Knight emphasised the influence of individual enthusiasts close to the government and the Admiralty, especially, in the earlier period, the president of the Royal Society, Sir Joseph Banks, and the second secretary of the Admiralty, Sir John Barrow[38]. In fact, there was a lot of support for science within the Admiralty and the naval administration in general. A succession of Admiralty officials, from first lords down to lowly clerks, were imbued with degrees of enthusiasm for science ranging from indifference, through benign well-wishing, to active participation and encouragement. After 1829, Sir Francis Beaufort in particular, a node in more networks than just the 'Cambridge Network',[39] acted as a major link, coordinating the activities of his survey vessels with the investigations of his scientific friends in London, Oxford, Cambridge, Glasgow, Dublin and around the world. In his day, Beaufort was almost certainly the member of the naval administration who consistently did the most to promote science in the Navy, but his allies included Barrow, first lords including Lord Melville, the Earl of Auckland, Sir James Graham, the Earl of Minto, and the Duke of Northumberland, and a host of lesser officials sympathetic to science, including people such as James Stephens, a clerk at the Admiralty from 1807 to 1845 who was a keen and knowledgeable entomologist, ornithologist and zoologist, who in 1818 was granted extended leave to work at the British Museum arranging and cataloguing the insect collection. Or Thomas Crofton Croker, another Admiralty clerk from 1818 to 1850, a leading light in the Society of Antiquaries, a founder member of the British Archaeological Association, and a horticulturalist. Or John and William Cresswell, George Smith and William Evans – all Admiralty clerks and all entomologists or horticulturalists or both.[40]

Opportunities to Pursue a Scientific Career in the Navy

Despite the tendentious context in which it appeared, there is little doubt that Charles Babbage's oft-quoted jibe, that 'the pursuit of science does not, in England, constitute a distinct profession', accurately reflected the situation around 1830.[41] Without

[38] Knight (1975).
[39] David Philip Miller, 'The Revival of the Physical Sciences in Britain, 1815–1840', *OSIRIS* (1986), **2**, 107–34, 107; Cannon (1978), 29–72.
[40] Cock (2003).
[41] Babbage (1830), 10.

pre-empting the discussion of 'professionalisation' below, or getting bogged down in definitions of potentially problematic terms such as 'professional', 'scientist' and 'career', it must be safe to assert that before the later nineteenth century there were few professional scientists in Britain and few opportunities to pursue a career in science.[42] The expansion of scientific activity in the Navy following the Napoleonic Wars, however, provided distinct opportunities for individuals in any of several categories to make at least some kind of scientific career for themselves either within, or in association with, the Navy.

Firstly, there were the many civilians who accompanied naval survey vessels and voyages of exploration to the Arctic, Congo and Niger, to act as naturalists, astronomers or antiquarian scholars. Charles Darwin must be the most famous member of this group: un-salaried, un-officially even a naturalist, and making only one, albeit five-year round-the-world, voyage, he did not exactly have a *career* in the Navy. Nor did others such as the biogeographer Hewett Cottrell Watson, or the naturalists and antiquarians Robert Pashley, Paul Wilhelm Forchhammer and Edward Forbes, who travelled in other naval vessels on somewhat similar terms. But the experience they gained certainly laid the foundation for their subsequent scientific careers. (Watson, already more established in his discipline than these others before his cruise to the Azores in the *Styx* in 1842, nevertheless made much capital from his trip.[43]) Other civilians, though, *were* paid salaries by the Navy: men such as the geologist Joseph Beete Jukes, the naturalists Allan Cunningham, George Tradescant Lay, Thomas Edmondston and Berthold Seemann, and John MacGillivray, who was employed as naturalist on three naval surveys between 1842 and 1855.[44] Yet others, including David Lockhart in the *Congo*, George Barclay in the *Sulphur* and J.W. Hamilton in the *Acheron*, were paid their salaries not by the Navy but by Kew Gardens, the Horticultural Society, or the New Zealand Company.

Another group of salaried naturalists who sailed in many naval vessels were the naval surgeons and assistant surgeons. Joseph Dalton Hooker and Thomas Henry Huxley were the two in this period who subsequently became most famous, but quite a number of naturalists started careers and in many cases pursued them over a long term, as surgeons in the Navy, including men such as the explorer, ichthyologist and

---

42 D.E. Allen, 'The Early Professionals in British Natural History', in Alwyne Wheeler and James H. Price, eds, *From Linnaeus to Darwin. Commentaries on the history of biology and geology*, London, 1985, 1–12; J.B. Morrell, 'Individualism and the Structure of British Science in 1830', *Historical Studies in the Physical Sciences* (1971), **3**, 183–204.

43 Frank N. Egerton, *Hewett Cottrell Watson. Victorian plant ecologist and evolutionist*, Aldershot, 2003.

44 On the first voyage, MacGillivray was employed as the private collector of the Earl of Derby; on the second and third he was the official naturalist, paid by the Admiralty (Robert Ralph, 'John MacGillivray – his Life and Work', *Archives of Natural History* (1993), **20**, 185–95).

all-round naturalist Sir John Richardson, the conchologist Arthur Adams, George Busk, Richard Brinsley Hinds, David Lyall, Robert McCormick, Benjamin Bynoe, William Webster, John Robertson, Charles Forbes, William Balfour Baikie, John Joliffe, Alexander Collie, Alexander Fisher, Harry Goodsir, Richard King, James McBain, Peter Cormac Sutherland, Thomas Thomson and a good many others.[45] By the 1840s, if not before, it was regarded at the Admiralty as a lost opportunity if a vessel going to any lesser-known part of the world were not accompanied by a competent naturalist; and the director-general of the Medical Department of the Navy, Sir William Burnett, and the Hydrographer, Sir Francis Beaufort, both made it a point of some importance to ensure that such vessels had on board either a civilian naturalist or a surgeon appointed specifically for his knowledge of natural history.

The third group who pursued a scientific career in the Navy were naval officers themselves. Many officers dabbled in science in spare moments during long voyages, and such amateur enthusiasms were often put to good use by Beaufort especially, who encouraged them to direct their attention to specific aspects of meteorology, tides, natural history, or whatever 'facts … present[ed] themselves'.[46] But there emerged in this period a group of officers who *specialised* in scientific work, in some cases making whole careers from doing little or nothing else. Sir Edward Parry and James Clark Ross specialised in scientific exploration. Parry acted as a Royal Society-appointed observer on the three expeditions he commanded to Arctic Canada, and an attempt on the North Pole in 1827.[47] Ross, with even greater polar experience, became an acknowledged expert on geomagnetism. Henry Foster was an expert on geodesy and magnetism. Frederick Beechey, William Hewett and John Washington specialised in tides, William Allen in scientific exploration in Africa, and Robert FitzRoy in tides, meteorology and other physical sciences. W.H. Smyth, Thomas Graves and Thomas Spratt became authorities on classical antiquities (appreciation of Spratt's work on surveying the ruins of the Mausoleum at Halicarnassus, for example, earned him a recommendation of promotion from the Trustees of the British Museum;[48] while Smyth also became a major figure in astronomy after his retirement, vice-president of the Royal Society, and president of the Royal Astronomical and Geographical Societies). Phillip Parker King was a naturalist; Sir Edward Belcher, Owen Stanley, Bartholomew Sulivan, Alexander Bridport Becher and Henry Bayfield were active in many areas. Some of these men obtained appointments *because of* their scientific

---

45 Cock (2003).

46 Hydrographic Office, Taunton, (HO) Minute Book 2, 246, Hydrographer's Instructions to Surveyors, Beaufort, 19 December 1835.

47 Michael T. Bravo, 'Science and Discovery in the Admiralty Voyages to the Arctic Regions in Search of a North-West Passage (1815–25)' (PhD diss., University of Cambridge, 1992).

48 HO In-Letters Prior to 1857, N206, Newton to Beaufort, 10 February 1848; Letter Book 15, 131, Beaufort to Forshall, 19 February 1848.

activities.[49] There were even cases of officers, such as Lt James Wood, and Captains Francis Price Blackwood and Sir Everard Home, blatantly soliciting appointments from the Admiralty Board on the strength of their scientific records and skill in using precision instruments.[50]

Also associated with the Navy for a time, though more loosely than these three groups, were Thomas Young, Michael Faraday and William Wollaston, who were paid salaries as commissioners on the Board of Longitude, between 1818 and the abolition of the board in 1828, a circumstance which Barrow recalled as 'the first attempt, that I remember, to open salaried office to men of science'.[51]

The Professionalisation of Science

Given that the word 'professional' carries connotations of competence and diligence as well as of paid employment, technical training, qualifications, organisation and the various other characteristics that sociologists and historians have sought to associate with it, and that many of the greatest figures in science through the ages have been people who by any satisfactory definition of the term could not be considered 'professional scientists', most scholars have lost faith in simplistic models of the process of 'professionalisation', and have suggested that it might be more profitable to focus instead on those who pursued careers in science, irrespective of how they fit retrospectively into such artificial categories.[52] As sketched out above, the Navy offered opportunities for many men to pursue at least some sort of scientific career, a circumstance that David Knight had, in his 1974 paper, interpreted as acting counter to professionalisation.[53] However, it is also possible to interpret the participation of naval officers and surgeons in science as tending to *promote* that process.

Describing the process by which science became professionalised, insofar as it did, in the nineteenth century, Jack Morrell drew together some general trends under six heads.[54] But he confined his account largely to the role of the universities, and it may be as well to see how the experience of the wider scientific community, and specifically the Navy, fitted his model. Firstly, he pointed to the increase in the number

---

[49] Some officers in the French Navy, too, achieved promotion on the strength of their scientific work (N.A.M. Rodger, 'Navies and the Enlightenment', in van der Merwe [forthcoming]).
[50] HO Misc.22, Folder II, W14, May 1850; HO In-Letters H859, 23 January 1832; HO In-Letters B42, n.d. *c.*1841.
[51] Knight (1975), 63.
[52] Cannon (1978), 137–200; Porter (1978), 823.
[53] Knight (1975), 58.
[54] J.B. Morrell, 'Professionalisation', in R.C. Olby, G. Cantor, J.R.R. Christie and M.J.S. Hodge, eds, *Companion to the History of Modern Science*, London and New York, 1990, 980–89.

of paid positions available to scientists, citing the increase in the number of university professorships in scientific subjects in France, Germany and Britain – which in the latter case doubled between 1820 and 1850. But it should also be noted, and it is more to the point since universities at that time were not primarily about *research*,[55] that there was in the same period a big increase in the number of scientific jobs, paid and unpaid, in the Navy. Scientific specialists also enjoyed, generally, a deserved status in the Navy, which was an organisation of experts who respected expert knowledge and skills.

Secondly, Morrell noted the introduction of specialist scientific qualifications. Before the new-fangled PhD system spread from Germany, and Oxford and Cambridge began awarding science BAs, later in the nineteenth century,[56] scientific training in Britain was organised more on the model of the apprenticeship, whether the trainee was Michael Faraday in Davy's Royal Institution laboratory, a member of the Geological Survey, an observer in Airy's Royal Observatory, or a scientific serviceman. That was one reason for the perpetuation of the sub-professional standing of researchers. Practitioners of medicine, on the other hand, had successfully asserted their professional status by the seventeenth century, and before the institution of specific scientific degrees, a medical degree was recognised as a general scientific qualification. But notwithstanding the transmogrification of the ancient Guild of Barber-Surgeons into the Royal College, the surgeon, lacking the theoretical doctrine, university education and status of the physician, ranked below him.[57] Although, or perhaps because, the pay and status of naval surgeons was, until the mid-nineteenth century, below that of army surgeons and many of those in private practice, the Navy was a focus of efforts in the 1830s to improve the standing of surgeons, especially junior ones, orchestrated in part by the radical MP for Finsbury and founder-editor of the *Lancet*, Thomas Wakley.[58] The Navy was at the van of developments introducing a foundation of university study leading to a formal qualification for its surgeons. At the start of the nineteenth century, the surgeon's mate, trained on the job, was phased out and replaced by the assistant surgeon, who was required to have completed a course of study at an approved university before joining the Navy, and who, if he had not done so already, was routinely granted periods of study leave to return to university and

---

55 Harvey W. Becher, 'Voluntary Science in Nineteenth-Century Cambridge University to the 1850s', *British Journal for the History of Science* (1986), **19**, 57–87; Hays (1983), 91–119.

56 Morrell (1990), 983.

57 A vestige of this inferiority is fossilised today in our hospitals, where surgeons (who traditionally lacked MDs) are still titled 'Mister', although now, of course, in an ironic twist, it denotes *higher* standing than that of the lowly 'Doctor'.

58 Christopher Lloyd and Jack L.S. Coulter, *Medicine and the Navy 1200–1900*, 4 vols, Edinburgh and London, 1957–63, **4**, 11–25; Sir Peter Froggatt, 'John Snow, Thomas Wakley, and *The Lancet*', *Anaesthesia* (2002), **57**, 667–75.

qualify as an MD.[59] Medical training invariably included a course in botany, in order that the surgeon or physician be able to identify plants with medicinal properties, and at Edinburgh's Medical School, for example, students were given instruction in how to collect scientific data in the field because their professors knew they might seek a military or naval career.[60] Since natural history and natural philosophy were explicitly specified in the regulations for entry into the Navy as an assistant surgeon as subjects of *particular* merit,[61] not only were naturalists and scientists in general encouraged into the Navy, but the professionalisation of science was able to hitch a ride on the back of the professionalisation of naval medicine. The standing of those practising medicine within the Navy steadily advanced through the nineteenth century: from mates and warrant officers at the end of the eighteenth century, to assistant surgeons as warrant officers and surgeons as commissioned officers, to even assistant surgeons obtaining commissions (in 1843), and then in 1873 to these highly educated and qualified professionals shedding the demeaning caveat 'assistant' to become full surgeons immediately on entry, while their superiors were given the title of staff surgeon.[62] And as their standing rose, so too did the status of the science so many of them were pursuing either as subsidiary, or in not a few cases, primary activities. Training for other classes of naval scientist, however, as for seamanship, continued to be largely by example and practical experience, and the Navy did not itself introduce specific scientific qualifications for its officers or offer them official study leave in the way it did its medical officers.

Morrell's third point was that scientific training was improved, especially through the development of the university laboratory. Much of the science done in the Navy required no laboratory; for the rest, one could be provided on board.[63] From Cook's *Endeavour* onwards (although perhaps it should be taken from Bougainville, who beat him by a year in this) whole naval vessels have been described as 'floating laboratories'.[64] On the many voyages to the Arctic in search of a North-West Passage, before low temperatures could be produced in the laboratory, the Navy used the Arctic as a low-temperature laboratory to investigate phenomena such as the properties of

---

59 PRO ADM 104/23–24 Assistant Surgeons' Records of Service, 1839–48.

60 Ex inf. David Knight.

61 PRO ADM 105/73, 135 and enclosure: 'Alterations to the Regulations for the Guidance of Candidates for Admission into the Naval Medical Service', Sir William Burnett, 1839 (rev. to 1842).

62 Lloyd and Coulter (1957–63), **3**, 32–4, **4**, 5–7; N.A.M. Rodger, *Naval Records for Genealogists*, Kew, 1998, 23.

63 This development reached its apogee, perhaps, in the *Challenger* expedition of 1872–76.

64 Glyn Williams, 'The Achievement of the English Voyages', in Derek Howse, ed., *Background to Discovery. Pacific exploration from Dampier to Cook*, Berkeley, 1990; Richard Sorrenson, 'The Ship as a Scientific Instrument in the Eighteenth Century', *OSIRIS* (1996), **11**, 221–36.

materials and the propagation of sound in temperatures down to minus 70° F.[65] And because naval science, like seamanship and navigation, was taught largely through practical experience, these open-air and ship-board laboratories promoted the training of the midshipmen, junior officers and, in some cases, seamen who were to become the next generation of scientific servicemen, just as much as the laboratories of Göttingen and University College.

The Navy may have had relatively little to do with the fourth aspect of professionalisation – rapid specialisation as esoteric technical languages developed and the various sciences became demarcated as areas of specific knowledge and expertise (but then, in most spheres, this development was not far advanced until the later years of the century). Although the Navy developed, and almost created in some cases, the sciences of tidology, meteorology, ocean currents and geomagnetism, many of the Navy's 'specialists' in these fields, up to the 1850s at any rate, remained generalists: F.W. Beechey, an expert on tides, was also an expert on meteorology; J.C. Ross, Edward Sabine's equal on geomagnetism, was a competent naturalist and was selected to lead a major tidal survey of the Atlantic; Robert FitzRoy, geologist, naturalist and tidal observer and theorist, applied to run the Compass Department and became the first head of the Meteorological Office.

Although their other professional loyalties, as officers of the Navy, might have presented complications, in practice the Navy can be seen also to have contributed to Morrell's fifth point about the professionalisation of science – the development of group-solidarity and self-consciousness among scientists – by playing such a large part in some of the scientific societies. Up to 1855, there were, for example, 181 naval officers, surgeons, administrators and civilian scientists closely associated with the Navy in the Royal Geographical Society, 86 in the Royal Astronomical Society, and 85 in the Royal Society.[66] Without that support, some of those societies (especially the RGS, and perhaps the RAS too) might not have been established in the first place, or would at best have had a lower profile and less access to government. Although there was a relatively small naval presence in BAAS (35 members), those there included heavyweights such as J.C. Ross, F.W. Beechey, George Back and John Washington.

Finally, Morrell cited the development of reward systems, including the award of honours, medals and FRSs for scientific achievement. Whether or not this was not equally a part of 'non-professional' science, it was another thing in which the Navy took a hand: Arctic and Polar medals and knighthoods were awarded to her scientific explorers;[67] as mentioned, several naval officers achieved the coveted FRS; and in

---

[65] Bravo (1992); W.E. Parry, *Journal of a Voyage for the Discovery of a North-West Passage ... 1819–20*, London, 1821.

[66] Cock (2003), Appendix: 'Scientific Servicemen in the Royal Navy: Scientific credentials'.

[67] Lt-Col. Neville W. Poulsom and Rr-Adm. J.A.L. Myres, *British Polar Exploration and Research. A historical and medallic record with biographies, 1818–1999*, London, 2000.

1848 the Admiralty Board issued an official memorandum holding out the prospect of 'pecuniary reward or promotion' to such of their officers and surgeons as did good scientific work.[68]

Conclusion

Partly *because* of its traditionally amateur standing, science was largely a gentlemanly pursuit,[69] and therefore a fitting 'occupation' for a gentleman-officer: whilst few made a living at science, there was little chance of being mistaken for a vulgar tradesman or working man – a 'mere collector' in the words of William Jackson Hooker.[70] In the period between the Napoleonic and Crimean Wars, there were many opportunities for individuals to do scientific work in or with the Navy, and some were able to specialise in aspects of this work to such an extent that they in effect made scientific careers for themselves. The question remains, whether these scientific servicemen should be regarded as *professional* scientists. They certainly seem to blur the distinction between, on the one hand, the professional salaried scientist in his laboratory, of whom perhaps Michael Faraday was the archetype, and on the other, the country parson indulging an amateur curiosity in botany. The naval officer was certainly a salaried professional; if a professional scientist is an expert paid to do science, then it depends what exactly a naval officer's job was, whether or not he was a professional scientist. It is clear from their actions and from the orders and instructions issued by the Admiralty, that it was generally accepted, at the time, that the job naval officers were paid to do (especially naval surveyors and surgeons) included science; indeed, in the case of surveyors, actually *was* science. Therefore, the scientific specialists among naval personnel, in many cases, have to be recognised as professional scientists.

A more important point, however, is what was actually going on in the first half of the nineteenth century, when the Royal Navy was heavily and deeply involved, from the Admiralty Board down to the midshipmen and, in some cases, the seamen, in promoting, supporting, organising, executing and publishing the results of, investigations into many of the major scientific questions of the day. The Navy was itself a relatively new profession, dating from the late seventeenth or early eighteenth century when 'gentlemen-officers', often lacking experience at sea, and skilled but ungentlemanly 'tarpaulins' had been replaced by a new class of formally-qualified professional naval officer.[71] At the beginning of the nineteenth century, the Navy was

---

[68] The memorandum was printed in the front of Herschel (1851).

[69] Cannon (1978), 145–6.

[70] HO In H894, Hooker to Beaufort, 26 March 1842.

[71] J.D. Davies, *Gentlemen and Tarpaulins: the Officers and men of the Restoration Navy*, Oxford, 1991.

the 'most modern and intellectually adventurous of professions', receptive to the idea that 'understanding would come through the amassing of facts',[72] and it can be seen as an incubator for professional scientists, and even for the scientific profession itself. Science was being professionalised by the Navy in those years, and largely at State expense, but this process was hidden from view *because* it was being done in the Navy, by officers who were already paid salaries as members of another profession. Thus, in its early stages, the scientific profession developed partly within the career-structure of another profession, that of the naval officer.

Any account phrased in terms of a process such as professionalisation is open to the objection that it smacks of that cardinal sin for the historian: Whiggishness. Leaving aside, therefore, the question of whether developments in the working lives of practitioners of science were really all tending to the same end, and also all consideration of whether or not it was a *good thing*, nevertheless it cannot be denied that by the end of the nineteenth century there were many more professional scientists in Britain than there had been at its beginning. The Navy, as one of the principal sponsors of science in the first half of the nineteenth century, played an important part in bringing that about, and, however we may choose to categorise them whilst they were serving in that organisation, there can be no doubt that the Navy provided the springboard for young men such as T.H. Huxley and others to jump into scientific careers as, recognisably, *professional* scientists.

---

[72] N.A.M. Rodger, 'Medicine and Science in the British Navy of the Eighteenth Century', in Christian Buchet, ed., *L'Homme, La Santé et La Mer. Actes du Colloque international tenu à l'Institut Catholique de Paris, les 5 et 6 décémbre 1995*, Paris, 1997, 333–44, 339.

Chapter 8

Expertise and Christianity: High Standards *Versus* the Free Market in Popular Publishing

Aileen Fyfe

Introduction

In 1853, the Reverend Thomas Pearson, a North Berwickshire nonconformist, remarked that, 'The age in which we live, is unprecedented for the cheapness and abundant supply of its literature'.[1] The success of the first penny periodicals, steam-printed in tens of thousands of copies in the 1830s, had encouraged book publishers to join the experiment of lower-priced reading material.[2] The widening availability of cheap print meant that readers in the working classes, who had previously had little or no access to newspapers, periodicals or books, began to have the opportunity to read either for entertainment, or for instruction in everything from politics to philosophy and chemistry.[3] The *North British Review* had commented a few years earlier that, 'it is the glory of our age to have brought science and sound literature within the reach of the humblest citizen'.[4] Pearson, too, paid homage to the great work done by the printing press, to which 'we greatly owe our civil and religious liberties', but he was worried that the products of the cheap press were not all religious or instructive. Rather, '[s]peculations, decidedly

---

[1] Thomas Pearson, *Infidelity: its Aspects, causes and agencies; being the prize essay of the British Organization of the Evangelical Alliance*, London, 1853, 478.
[2] John Feather, *A History of British Publishing*, London, 1988; Patricia Anderson, *The Printed Image and the Transformation of Popular Culture 1790–1860*, Oxford, 1991; John O. Jordan, and Robert L. Patten, eds, *Literature in the Market Place: Nineteenth-century British publishing and reading practices*, Cambridge, 1995.
[3] Richard D. Altick, *The English Common Reader; a Social history of the mass reading public, 1800–1900*, London, 1957; Victor E. Neuburg, *Popular Literature: a History and a Guide*, Harmondsworth, 1977; James Raven, Helen Small, and Naomi Tadmor, eds, *The Practice and Representation of Reading in England*, Cambridge, 1996; Jonathan Rose, *The Intellectual Life of the British Working Classes*, New Haven, 2001; David Vincent, *Bread, Knowledge and Freedom: a Study of nineteenth-century working class autobiography*, London, 1982.
[4] [Coventry Patmore], 'Popular Serial Literature', *North British Review* (1847), 7, 110–36, 124.

hostile to true religion and to man's best interests ... have descended through the many channels opened up by the prolific press, to the reading millions'.[5]

Addressing an audience at the Society of Arts Educational Exhibition in London in 1854, Cardinal Nicholas Wiseman described cheap publications as 'reptiles' which crawl 'in their own slime on the surface of the earth, and like the venomous serpent at the beginning of the world, insinuate themselves into the peaceful and happy domestic circle, and there introduce pain, and ruin, and death, and taint the whole of a rising family'.[6] Wiseman had been appointed the first archbishop of Westminster only a few years earlier, and evangelicals like Pearson had certainly not forgotten the consequent outcry of 'Papal Aggression'.[7] Yet, on the dangers of cheap publishing, voices from across the political and religious spectrum found themselves, for once, agreeing with the cardinal.[8] Instead of 'intellectual food ... of a wholesome quality', cheap publications were providing the working classes with 'mental poison'.[9] On one level, there was a problem with what the nonconformist *British Quarterly Review* called 'that express literature of trash and garbage', in which the fiction was at least 'highly seasoned', and the non-fiction was nothing but 'unmitigated platitude' and, too often, 'error and superstition'.[10] Such material could provide no lasting benefit to its readers. Even more worrying was the 'flood of impure and anti-social works' which were 'immoral and soul-destructive' and 'of a most debasing and pernicious character'.[11] Pearson's comments had been made in the context of his prize-winning essay on *Infidelity: its Aspects, causes and agencies*, which named the press as one of the major causes of the spread of religious unbelief.

Cheap publications had the potential to be dangerous because they were being read by significantly more people, further down the socio-economic scale. A typical, ten-shilling-per-volume book would reach perhaps a thousand or so educated middle-class readers, but the penny periodicals had sales figures in the tens of thousands, and by the 1840s, were reaching hundreds of thousands.[12] Books were never as cheap

---

[5] Pearson (1853), 477, 478.

[6] Wiseman, 'Home Education of the Poor', quoted in [David Masson], 'Present Aspects and Tendencies of Literature', *British Quarterly Review* (1855), **21**, 157–81, 178.

[7] Evangelical responses to the restoration of the hierarchy, and other events of the late 1840s and early 1850s, are discussed in Donald M. Lewis, *Lighten their Darkness: the Evangelical mission to working-class London, 1828–1860*, London, 1986, ch. 8.

[8] Morag Shiach, *Discourse on Popular Culture: Class, gender and history in cultural analysis, 1730 to the present*, Cambridge, 1989, 73–4, discusses these concerns.

[9] Mrs Percy Sinnett, 'What is Popular Literature?', *The People's Journal* (1848), **5**, 7–8, 8; Letter to the editor by the author of *The Power of the Press*, *British Banner* 07/01/1848, 15.

[10] [David Masson] (1855), **21**, 157–81, 170, 174.

[11] Address by the committee of Religious Tract Society, quoted in *British Banner* 21/06/1848, 438; Religious Tract Society (hereafter RTS) Annual Report (1851), 123.

[12] Anderson (1991), chs 3–5.

as periodicals, but once prices came down to a shilling or two at most, audiences increased significantly, as was demonstrated by George Combe's *Constitution of Man* (1828), whose 1*s*.6*d*. People's Edition sold almost 7,500 copies in 1835, and a further 35,000 the following year.[13] Books or periodicals which were cheap enough to reach these enormous audiences were described as 'popular', for they seemed to be reaching 'the people'. At mid-century, the connotations of the term 'popular' became more complex, partly as a consequence of changing conceptions of 'the people' and the 'mass' audience.[14] In particular, in relation to literature, 'popular' came to mean a style of expository writing, which dealt with abstruse material in a non-technical and easy to understand manner.[15] When contemporaries referred to 'popular treatises and essays' they increasingly meant treatises which were adapted for a wide audience (the 'people') in terms of both price and assumed educational background.[16] 'Popular' did not initially carry derogatory implications about the authors' lack of expertise, or simplifications or distortions of their material, as the modern usage (a 'mere popular work') would imply, but as the 1840s and 1850s wore on, it became increasingly opposed to genuine, expert knowledge.

For historians of science, the best-known example of a work which received heavy criticism around this time is undoubtedly the *Vestiges of the Natural History of Creation* (1844).[17] It is only in retrospect, however, that we can label *Vestiges* a 'popular' work. It was not typical of the sorts of books which so alarmed Pearson and Wiseman, even though it was criticised for its author's racy literary style and for its lack of religion. Its first edition cost 7*s*.6*d*. and amounted to just 750 copies, and, although its author was anonymous, he laid claim to expertise in advancing his new theory of cosmic development.[18] As James Secord has argued, this combination of features made it difficult for contemporaries to decide where to locate *Vestiges*.[19] By price and initial readership, it was not 'popular'. In its attempts to contribute to scientific debates, it was not simply expository. It was only after concerted efforts by

---

[13] John M. van Wyhe, 'Phrenology's Nature and the Spread of Popular Naturalism in Britain, *c*.1800–1850' (PhD diss., University of Cambridge, 2001), appendix C.

[14] On popular literature, see Neuburg (1977). On 'popular', see Shiach (1989), introduction (esp. 32–3) and ch. 3. On the mass audience, see Jon P. Klancher, *The Making of English Reading Audiences, 1790–1832*, Madison, 1987, ch. 3.

[15] The *OED*'s first record of 'popular science' was in 1841, see 'popular' adj. meaning 4.a. The associated verb 'to popularise' had appeared in 1833.

[16] [Masson] (1855), 166.

[17] Richard Yeo, 'Science and Intellectual Authority in Mid-Nineteenth Century Britain: Robert Chambers and *Vestiges of the Natural History of Creation*', *Victorian Studies* (1984), **28**, 5–31; James A. Secord, *Victorian Sensation: the Extraordinary publication, reception and secret authorship of Vestiges of the Natural History of Creation*, Chicago, 2000.

[18] Secord (2000), 131.

[19] *Ibid*., 17–24.

men of science to establish both their own authority and the Vestigian's lack thereof that *Vestiges* eventually stood condemned as 'popular' for its lack of expertise.

This essay examines the debate surrounding popular literature in the 1840s and 1850s, for it is only by understanding this broader debate that historians of science can appreciate what it meant to describe works of science as 'popular'. Several extended commentaries on popular literature appeared in the literary reviews, and they indicate the level of concern felt about the phenomenon. The commentators were torn. They could not doubt that cheap, expository works of non-fiction were desirable, necessary and worthwhile, and its writers were to be commended for assisting the cause of education and self-improvement. (This support certainly did not extend to cheap fiction, which was, as Pearson put it, 'calculated to make men and women any thing but wise and thoughtful'.[20]) Unfortunately, too many cheap expository works failed to live up to expectations, and commentators routinely complained of the writers' lack of ability, and their tendency to disregard the need for a proper Christian tone. The *British Quarterly Review*, for instance, maintained that the editorial work of most of the periodicals which claimed to provide instruction was 'executed by a pair of scissors, with an incredibly small amount of intelligence to guide them'.[21] Commentators agreed that publications which were sloppily produced and full of errors were hardly desirable things for the working classes to be reading, and, as we shall see, they felt that this was a problem that could be solved by a careful selection of writers, and a commitment to literary standards. There was far less common ground among the commentators on what counted as an appropriate religious tone. Most agreed that overtly anti-Christian works should be condemned, but there was disagreement over the amount of explicit Christianity that was desirable. Running unanswered throughout the debate was the question of why so many readers kept buying works which reviewers thought were clearly and obviously inferior. The tastes of the mass reading public proved an obstacle to all plans for improving the state of cheap expository literature.

The Competence of Authors

In 1855, the *British Quarterly Review* discussed the type of authorship required to produce the current diversity of 'popular treatises and essays without number, and on all subjects – geology, political economy, politics and whatnot'.[22] The reviewer was David Masson, himself an experienced writer of such treatises for the respected

[20] Pearson (1853), 501.
[21] 'Cheap Literature', *British Quarterly Review* (1859), **29**, 313–45, quotation on 330–31.
[22] [Masson] (1855), 166.

Edinburgh educational publishers, W. & R. Chambers. Masson characterised the aim of 'expository authorship' as 'working down the truths and generalities of the various sciences to the apprehension of the public'.[23] He claimed that the key requirement was what we might term *literary* ability – the ability to write in a suitable style, explaining complex issues in a readable, non-technical manner. It did require a firm grasp of the specific subject matter, but there was no need for extensive expertise. All the expository writer had to do was 'read conscientiously certain authorities on the subject on which he is to exercise his craft and reproduce their matter in his own form'.[24] This form of authorship was distinct from what Masson labelled 'high, or true' authorship, which involved a 'labour of research, [and] ... examination of the original materials by and for himself'.[25] Although undoubtedly 'useful and honorable' work, expository writing was not usually the preserve of 'men of truly superior faculty'.[26] It should rather be undertaken by an 'intelligent and conscientious workman'.[27]

The status of such a 'workman' is apparent in the case of William Martin, a former employee of the Zoological Society who made his living in the 1840s and 1850s writing expository works of natural history for Charles Knight, and the Religious Tract Society. Martin had originally hoped to make his name as a man of science, but when the comparative anatomist Richard Owen wrote him a letter of reference, he described Martin as having been 'most industriously and honorably occupied in diffusing sound scientific information, in Zoology, amongst the general and youthful readers of the English language'.[28] Owen recognised the utility and necessity of Martin's work, but his choice of 'industrious and honorable' adjectives, with the insistence that Martin's ability lay in the 'diffusion' of information, made clear that Martin was a useful and talented workman, not an original scholar. Even Martin himself came to take this view of his writings, referring to them in 1853 as being 'not brilliant' and having a 'plain utility'.[29]

Masson acknowledged that more superior authors might occasionally engage in expository authorship, but implied this was a rare event. John Crosse, in the *Westminster Review*, had admitted as much, *a propos* of *Vestiges*, when he remarked that the men of

---

[23] *Ibid.*, 166.
[24] *Ibid.*, 163.
[25] *Ibid.*, 164. Most scholarly studies of nineteenth-century authorship have been more interested in 'high' literature, particularly fiction, for instance, John Sutherland, *Victorian Novelists and Publishers*, London, 1976. For general studies, see Nigel Cross, *The Common Writer: Life in nineteenth-century Grub Street*, Cambridge, 1985, and Victor Bonham-Carter, *Authors by Profession*, London, 1978.
[26] [Masson] (1855), 167.
[27] *Ibid.*, 166.
[28] Royal Literary Fund (hereafter RLF) archives, item 1315.4, Owen to RLF 11 April 1853.
[29] RLF 1315.11, Martin to RLF, 16 April 1853.

science who would be 'competent to the task' generally 'disdain to popularise science'.[30] Moreover, he went on, the few exceptions who did write general works, such as John Herschel and William Whewell, were criticised for the high level of general education needed to appreciate their works. Their works might be appropriate for serious students, but they 'afford no broad and beaten track on which the multitude can travel onward, or food to satisfy the cravings of its unenlightened reason'.[31] For publishers hoping to sell cheap expository works by the tens of thousand, the level at which they were pitched was crucial. Low price was obviously essential, but it was insufficient. The Society for the Diffusion of Useful Knowledge (SDUK, 1826–46) had discovered this with its sixpenny treatises by expert authors, which failed to sell as well as anticipated, and found most of their audience among the middle classes. The society's competitor William Chambers later described them as 'on the whole too technical and abstruse for the mass of operatives'.[32] Publishers aiming at a mass market had to make sure that the language used in their works was appropriate to the educational level of their readers. With the exception of the rare few who were particularly skilled at writing in a non-technical style, experts were felt to be unsuitable writers of expository works.[33]

There was a further, more mundane reason for publishers' disinclination for expert authors. Until the later nineteenth century, authors were usually paid a one-off sum for the purchase of copyright in their work, with the publisher's valuation of the manuscript dependent on its length, the quality of the writing, the potential market, and the writer's reputation. Well-known expert authors would have to be paid more, which could have severe implications for publishers wishing to issue cheap works. Steam technologies had done much to reduce the cost of bulk production, leaving the cost of paper and of copyright as the major expenses. W. & R. Chambers were pioneers in the steamprinting of books, and, in 1837, they were able to produce 3,100 copies of their People's Edition reprint of William Paley's *Natural Theology* for only £113.[34] This

---

[30] [John Crosse], 'Review of *Kosmos* and *Vestiges*', *Westminster Review* (1845), **44**, 152–202, 153.

[31] *Ibid.*, 153. A similar criticism of men of science as popular writers was made by Edward Forbes, in the *Literary Gazette*, in 1851, see Secord (2000), 476.

[32] William Chambers, *Memoir of Robert Chambers with Autobiographical Reminiscences*, New York, 1872, 213.

[33] On the International Scientific Series (a later attempt to get expert men of science involved in writing for a wider audience), see Roy M. MacLeod, 'Evolutionism, Internationalism and Commercial Enterprise in Victorian Britain: the International Scientific Series 1871–1910', in A.J. Meadows, ed., *The Development of Science Publishing in Europe*, Amsterdam, 1980, 63–93 and Leslie Howsam, 'An Experiment with Science for the Nineteenth-Century Book Trade: the International Scientific Series', *British Journal for the History of Science* (2000), **33**, 187–207.

[34] Chambers Publications Ledger 1842–45, W. & R. Chambers archive, National Library of Scotland (hereafter NLS), Dep 341/274, 351.

allowed the books to be sold at the incredibly low price of 1*s*., and still bring in a profit of £42. The economic viability of such projects depended on the low production costs, and the absence of any payment to the author. Reprinting out-of-copyright works was financially sensible, but, by the 1840s, Chambers were increasingly commissioning new works. The firm was committed to the cause of working-class education, and had realised that works originally written for a relatively educated middle-class audience were difficult for working-class readers to follow. Commissioning works, of course, cost money. The expert writers of the SDUK's Library of Useful Knowledge had been paid around £180 per volume, and these were not unusual figures for works of that length.[35] However, payments of that order of magnitude would have forced Chambers to double or triple the retail price of their works. For their Educational Series, whose volumes ranged in price from 6*d*. to 4*s*., Chambers paid around £40 per work.[36] A similar situation existed at the Religious Tract Society (RTS), where authors for the sixpenny Monthly Volumes were paid around £30 each.[37] Even for their very longest works, neither of these publishers paid much over £120. Chambers and the RTS may have preferred not to use experts for fear of technical jargon, but, equally, they could not have published such affordable works if they had paid experts' fees.

The key asset of the successful expository author, therefore, was not subject expertise, but literary ability, and, as Charles Knight's praise of William Martin for 'his industry, punctuality, and conscientiousness' indicates, this included practical issues like meeting deadlines, as well as writing style.[38] Although publishers could find this ability in a wide range of people, school teachers were a favourite at Chambers, and ministers at the RTS.[39] School teachers and ministers were in the income bracket where even a small addition to a meagre salary would be greatly appreciated, and, with few pretensions to greatness, they were likely to obey publishers' stipulations about length, and accept suggested revisions. They could also be assumed to be educated and responsible, and although they might be lacking in specialist expertise, they knew how to use libraries. One of the RTS authors, for instance, was the Reverend John Kennedy, a Congregationalist minister in Stepney, East London. He wrote books and periodical articles on the River Jordan, Arabia, volcanoes, the search for Arctic explorer

---

[35] For more details of SDUK finances, see Scott Bennett, 'Revolutions in Thought: Serial publication and the mass market for reading', in J. Shattock and M. Wolff, eds, *The Victorian Periodical Press*, Leicester, 1982. John Murray paid the authors of his Family Library volumes around £220, see Scott Bennett, 'John Murray's Family Library and the Cheapening of Books in Early Nineteenth-Century Britain', *Studies in Bibliography* (1976), **29**, 139–66.

[36] For Chambers Educational Series, see Sondra Miley Cooney, 'Publishers for the People: W. &R. Chambers – the Early Years, 1832–50' (PhD diss., Ohio State University, 1970), ch. 4.

[37] For the RTS Monthly Series, see Aileen Fyfe, *Science and Salvation: Evangelicals and popular science publishing in Victorian Britain*, Chicago, 2004.

[38] RLF 1315.6, Knight to RLF, 12/4/1853.

Sir John Franklin, and the preacher Thomas Chalmers. He also wrote on ethnology and anthropology for John Cassell, another publisher of cheap educational works.[40] It is hardly surprising that a minister would feel capable of writing about the River Jordan or Arabia, but the works on the arctic, volcanoes and anthropology were rather less linked to Kennedy's religious vocation. Indeed, these must have been researched in his library (and that of the neighbouring ministerial training college). For Kennedy, research for his publications meant book-research, not original investigation. Ministers and teachers also had experience in making themselves understood to audiences with no specialist education, and perhaps little general education, whether in the classroom or from the pulpit. They were thus well placed to take specialist knowledge and turn it into generally understood language. They were also likely to be committed to the cause of education, and to see their role as a great responsibility. According to the *North British Review,* authorship was a most worthy vocation: 'It is no small thing ... to guide men to light from darkness, to truth from error'.[41] Even though their works were based on library research, such responsible authors could be trusted to use the right sources, to read the authoritative accounts, and not to rely on existing popularisations.

This shared commitment to high-quality instructive publishing was crucial to the projects of Chambers and the RTS, and its absence in the works produced by some other firms was one of the major reasons for complaints about the genre of expository publishing as a whole. Masson, in the *British Quarterly*, explained why book-based research by non-expert authors might be undesirable when he claimed that, if one were to take 'almost any three historical compilations, bearing no certificate in the name of their author ..., they will be found to be three different dilutions of some fourth compilation, which again, in its turn, may be the third dilution of some substantial book'.[42] Authors less responsible than John Kennedy might not bother to seek out the work of authorities, but might rely on some intermediary, and thus produce a new compilation with even less substance. As Masson implied, such works were often the result of hasty hack work, where the publisher was more interested in speed and cheapness, than in educational standards. The depressing thing for commentators was that these works seemed to sell just as well as the more reputable literature.

---

[39] On Chambers' authors, see Cooney (1970), 175–81. On RTS authors, see Fyfe (2004), chs 5–6.

[40] John Kennedy, *Old Highland Days: the Reminiscences of Dr John Kennedy, with a sketch of his later life by his son, Howard Angus Kennedy, with twenty-two portraits and illustrations*, London, 1901, 249–50.

[41] [J.W. Kaye], 'Pendennis: the Literary profession', *North British Review* (1850), **13**, 335–72, 371.

[42] [Masson] (1855), 170.

Religious Spirit

In his essay on *Infidelity*, Pearson had claimed that, 'In literature and science, we have not a little in which upper and under currents of scepticism are too perceptible'.[43] The sciences were often mentioned in this context, and Pearson particularly worried that a 'positive hostility to a pure spiritual religion, or that contemptuous disregard of it' had become 'wofully characteristic of some modern works of science'.[44] However, the sciences were certainly not felt to be the only areas contributing to the spread of atheism. Fiction was frequently seen as lacking Christian morality, while history, in the hands of Hume, Gibbon and their successors, had become 'the medium of insidiously communicating the poison of infidelity'.[45]

A pamphlet published in 1847, and written by an anonymous Congregationalist from Paddington, London, succeeded in rousing a reaction to the infidel press by quantifying the threat.[46] The author of *The Power of the Press; Is it rightly employed?* argued that although the press had originally been a gift from God, it was now being used by Satan in 'his open and unblushing attacks on truth, in his subtle and secret insinuations of error, to minister to the meanest passions or to the mightiest intellect'.[47] As evidence, the pamphleteer pointed to the large circulations of three 'obnoxiously irreligious and demoralising' weekly papers (presumably including the *Oracle of Reason* and its successor, the *Movement*), which achieved annual sales of 5.4 million in 1843. Since 1843, those papers had increased their circulations, and a further four weekly papers had appeared, with additional annual sales of around 5.7 million.[48] These new papers presumably included the *Family Herald*, the *London Journal*, and *Reynolds' Miscellany*, and although the pamphleteer admitted that the new papers were 'less debasing and less avowedly irreligious' than the earlier ones, he nevertheless regarded their influence as 'pernicious', since they 'practically desecrate the Christian Sabbath' and treat with 'disrespect every religious ordinance which the Christian would consider essential to the welfare of the community'.[49] The pamphleteer's initial figures for total annual circulations were 28.9 million for the irreligious press, against a paltry 3.8 million for the religious press. Even after revision, to include all the Bibles and tracts and periodicals printed in Britain, the religious total still came

---

43 Pearson (1853), 480; see also xiv.

44 *Ibid.*, 359.

45 Letter from Madras Committee, quoted in RTS Annual Report (1847), 116–17.

46 For the Paddington identification, see 'Letter to the editor, by the author of *The Power of the Press*', *British Banner* 07/01/1848, 15.

47 *The Power of the Press: Is it rightly employed?*, London, 1847, 5.

48 *Ibid.*, 7, 10.

49 *Ibid.*, 11. Patricia Anderson has argued that, nevertheless, these journals did inculcate many traditional moral values, see Anderson (1991), ch. 4.

to only 24.4 million.[50] Despite all the benefits that the press was clearly bringing to religion and education, it seemed all too clear that 'the demon of infidelity is stalking abroad' and infidelity is 'coming in like a flood'.[51]

Moreover, as the pamphleteer and Pearson both recognised, the avowedly infidel publications were only half the story.[52] There were also such periodicals as *Chambers's Edinburgh Journal* and the recently defunct *Penny Magazine* which had circulations as large as the infidel papers, and, despite being far more respectable, well-edited and well-produced journals, had contributors who (according to Pearson) 'would seemingly reckon it weakness, or fanaticism, to be indebted to [divine] revelation for a sentiment, a principle, or an embellishment', and so 'Christian truths and principles are ignored when they might have been most fittingly introduced'.[53] This was deeply dangerous, as it could lead readers to imagine that religion was not necessary – that secular knowledge about history, science or geography was all that was needed in this life. Coventry Patmore reminded the readers of the *North British Review* that a secular work 'is no neutral, but an enemy cruising under a neutral flag'.[54] Indeed, such works might cause more harm than overtly anti-Christian ones, for their veneer of respectability gave them access to homes which 'would justly repel the organ of an atheistic secularism'.[55] The anonymous pamphleteer was slightly more positive, admitting that the secular works sometimes contained 'a kind of striving after moral purity' or 'a seeking after sanctification (so to speak)'. However, since 'Christian atonement and justification are unknown', he was adamant that these journals were failing their readers spiritually.[56]

Commentators from across the religious spectrum could share the pamphleteer's distress about the rise of explicitly atheistic journals, but secular works, such as *Chambers's Edinburgh Journal*, were a particular dilemma for evangelicals. On the one hand, as the *North British Review* pointed out, Chambers' publications should be commended for their 'high standard … of taste, and tone, and feeling', and there would be a great general improvement if it were possible 'to elevate the moral tone of our lightest literature up to the present mark of Chambers'.[57] Even the anonymous pamphleteer had acknowledged that Chambers' publications were far superior in instructive and moral tone and production values to most of the other cheap publications available.[58]

---

[50] *Power of the Press* (1847), 15–17.
[51] Pearson (1853), xiv; *Power of the Press* (1847), 20.
[52] Pearson (1853), xiv.
[53] *Ibid.*, 480.
[54] [Patmore] (1847), 117.
[55] Pearson (1853), 503.
[56] *Power of the Press* (1847), 20.
[57] [Patmore] (1847), 124, 125.
[58] *Power of the Press* (1847), 20.

But, while liberal progressive commentators might find nothing amiss in the tone of Chambers's works, evangelicals could not ignore the fact that, as Pearson expressed it, 'In Chambers' publications, we miss the evangelical element – that decidedly Christian tone'.[59] If the circulations for these journals were added to those of the explicitly infidel, the pamphleteer's calculations made clear that the religious press trailed even further behind.

The pamphleteer (and his supporters at the newly founded *British Banner* weekly newspaper) regarded greater activity from the religious press as the solution, and particularly urged the biggest of the evangelical publishing organisations, the RTS, to do more.[60] The RTS was, however, already well aware of the problem caused by 'the rapid extension of secular information and the unexampled activity of the sceptical and licentious press', and had begun its response in 1845.[61] The relatively slower response of other sections of the religious press enabled one of the RTS supporters to claim in 1849 that it was 'the only instrumentality which is in operation at the present time ... with a view to the counteracting of an evil, and the diffusion of a good influence by means of literature'.[62] The RTS's Monthly Series (to which both John Kennedy and William Martin contributed) consisted of 100 volumes of non-fiction, selling at sixpence in paper and tenpence in boards. When the first volume (*Life of Julius Caesar*) sold out of its 10,000 copies, the society felt fully justified in its 'endeavour to diffuse literature and science in connexion with Scripture principles', and began to print in runs of 15,000.[63] These impressive sales figures suggested that well-written non-fiction with a Christian tone was a marketable product, even in the midst of so many competitors.

Discriminating Readers?

The *British Quarterly* had described how the French government, faced with a similar problem in the cheap press, had attempted to censor cheap publications, but for British commentators, this was not a realistic option, let alone one consonant with the prized

---

[59] Pearson (1853), 510.

[60] For the link between *Power of the Press* and the *British Banner*, see 'Prospectus' for the *British Banner* (1847) and 'Letter to the editor, by the author of *The Power of the Press*', *British Banner* 07/01/1848, 15.

[61] RTS Annual Report (1850), 141. On the RTS's secular publishing in the 1840s, see Fyfe (2004), especially chs 1–2. On evangelical efforts to reform the reading (and religious) habits of the working classes, see Lewis (1986), chs 5–6; R.H. Martin, *Evangelicals United: Ecumenical stirrings in pre-Victorian Britain 1795–1830*, Metuchen, NJ, 1983, sections III and IV.

[62] S.M. Peto, 'Address to RTS May meeting', quoted in *British Banner*, 16/5/1849, 306.

[63] RTS Annual Report (1846), 103.

concept of liberty.[64] This was why cardinal Wiseman had urged his listeners 'to provide to the utmost the antidote against their venom'.[65] Whether they were arguing for higher literary standards, or for more Christian tone, commentators agreed that the solution was to 'let good men employ a like energy in disseminating their cheap good things as bad men employ in disseminating their cheap bad things'.[66] If there was enough 'wholesome popular literature', it would eventually drown out the undesirable elements. Thus, reviews regularly praised those publishers – including Chambers and the RTS – who tried to uphold high standards in cheap expository works. Equally regularly, they expressed saddened amazement that the inferior works managed 'unhappily, [to] secure a ready sale, and liberally compensate the publishers'.[67]

The argument for increasing the free trade in cheap publications as a way of improving the cheap press was predicated on the assumptions that it was obvious which material was sound and which was potentially corrupting, and that readers who were fully aware of the various available options would choose to read wholesome popular literature.[68] This optimism about the mass reading public's desire for high-quality expository works must have begun to fade by the later 1850s. The sales figures tell the story. By the mid-1850s, the *London Journal* was selling 450,000 copies a week, despite being in the group condemned (by the *British Quarterly*) as 'foolish, frivolous, childish, and occasionally stupidly wicked'.[69] In comparison, the competing weekly periodicals from Chambers and the RTS were far behind. The RTS had launched a new weekly penny journal in 1852, called the *Leisure Hour*, which was praised by Pearson for 'combining instruction and entertainment, and … elevating the working classes in the scale of moral being'.[70] However, it sold around 60,000 copies a week at its launch in 1852, rising to almost 70,000 by 1855.[71] *Chambers's Journal* was selling 64,000 copies in the late 1840s.[72] As Chambers freely admitted, 'Our object all along has been to reach the masses, but we cannot get to them.'[73]

Despite its ambitions to counter the infidel as well as the secular press, the RTS was primarily effective against secular publishers such as Chambers. Like the periodicals, its books sold in the same sorts of numbers as Chambers', and, one assumes, to the

---

64 This affair, and the British reaction, is described in [Masson] (1855), 177–81.
65 Wiseman, 'Home Education of the Poor', quoted in *ibid.*, 178.
66 Pearson (1853), 509.
67 Address by the committee of RTS, quoted in *British Banner*, 21/6/1848, 438.
68 The most recent study of working-class reading habits is Rose (2001).
69 'Cheap Literature' (1859), 340. For *London Journal* circulation, see Anderson (1991), 90.
70 Pearson (1853), 509.
71 RTS Executive Committee Minutes, 02/03/1852; Finance Committee Minutes, 26/09/1855. RTS/USCL Archive, School of African and Oriental Studies, London.
72 Cooney (1970), 98.
73 'Booksellers', *Chambers's Journal*, ns 7, 06/02/1847, 88.

same sorts of people: the lower-middle classes and a few dedicated skilled artisans.[74] Neither publisher seemed to be able to reach what the *North British Review* described as a 'lower depth' in society, where the reading audience was much larger, and which was currently supplied 'with matter more congenial, by coarser and less scrupulous writers'.[75] Writing in 1858, Wilkie Collins suggested that the problem was that the mass reading public was too undiscriminating, and simply accepted the poor quality works along with the better ones. He argued that these readers 'are evidently, in the mass, from no fault of theirs, still ignorant of almost everything which is generally known and understood among readers whom circumstances have placed, socially and intellectually, in the rank above them'. They were 'waiting to be taught the difference between a good book and a bad', which, he hoped, was 'probably a matter of time only'. Eventually, the mass reading audience would 'learn to discriminate'.[76] And then the standard of cheap publications would have to be improved, for readers would vote with their purchasing power.

Yet, in reality, the mass public was already very discriminating. Its members just did not make the same choices that Collins or the quarterly reviewers would have made. Given the choice between the *Leisure Hour* and the *London Journal* – both illustrated penny weeklies – most preferred the *London Journal* every time. The *British Quarterly* urged the *Leisure Hour* to include the 'occasional sprinkling of lighter matter', as a sure way 'to attain the influence enjoyed by some of its inferior contemporaries', but there had to be a balance between the content needed to attract readers, and the standards needed to carry out the mission of improvement.[77] Despite their different religious attitudes, *Chambers's Journal* and the *Leisure Hour* shared ambitions for improving the working classes, and doing so through a well-produced publication with reliable content. Large numbers of working-class readers were clearly quite able to recognise this sort of journal, and they chose to avoid it.

It should be clear that worries about the expertise of authors and their religious world-views were hardly unique to *Vestiges* or to works of science, but were issues which had to be negotiated in relation to all cheap expository works. Publishers who were committed to the improvement of the working classes had a practical resolution to the problem of authorial competence. They were much more concerned to have authors with sound literary skills, able to do library research, and committed to producing a reliable and accurate publication, than to have the greatest authorities of the day. W. & R. Chambers and the RTS had much in common, with their shared commitment

---

[74] For book sales (ranging from a couple of thousand to the occasional 30,000), see Chambers Publications Ledger No. 2, NLS, Dep 341/275.

[75] [Patmore] (1847), 124.

[76] Wilkie Collins, 'The Unknown Public', *Household Words* (1858), **18**, 217–22, 222.

[77] 'Cheap Literature' (1859), 344.

to using the cheap press to produce high-quality instructive material, but they were at odds over the role religion should play in popular works.

In the mid-1850s, there was room for both congratulation and continuing concern among those who reviewed the state of cheap literature. A small number of reputable and committed publishers had made sure that cheap, non-technical books on a variety of general subjects were widely available. They may not have been written by the best experts in their fields, but they were reliable introductory works, thanks to the publishers' determination to choose responsible authors. Evangelical commentators may have regretted that this literature was not wholly Christianised, but could be pleased that there were plenty of Christian works available, and that many of the secular ones were respectable. The big problem continued to lie with the lowest reaches of cheap literature, what the *British Quarterly* had described as 'trash and garbage', and which included the pernicious and immoral, as well as the less harmful infinitely diluted compilations. As long as such works sold in large numbers, and as long as respectable publishers were unwilling to compromise their standards, the free trade in print continued to work against those wishing to insist upon either high authorial standards or explicit Christianity. The argument that good would win out over bad in a free fight seemed to be proven wrong, as readers voted with their pennies. The best that could be said about this region of literature was to hope that 'The taste for reading [being] once communicated to a new class of the community' would eventually lead them to 'a superior popular literature'.[78]

---

[78] [Masson] (1855), 181.

PART II
BELIEFS UNDERLYING SCIENCE

Chapter 9

Darwinian 'Becoming' and Early Nineteenth-Century Historiography: The Cases of Jules Michelet and Thomas Carlyle

Richard Somerset

This paper starts from an argument which I developed in a recent article relating to pre-Darwinian science and historiography specifically in France.[1] The article argued that in France there are stronger parallels with a Darwinian conception of 'becoming' to be found in the historiography of the period than in its natural science. The reasons for this tendency are essentially structural, relating in particular to the different impact of the early nineteenth-century shift towards an organicist mode of thought upon scientific discourse on the one hand and upon historical discourse on the other. Organicism was of course responsible for precipitating the change from taxonomical natural history to biology – the study of life as a particular variety of natural phenomenon. This shift also made possible the historicisation of the Chain of Being, and therefore opened the way for transmutationist models such as Lamarck's. However, this historicisation tended to remain superficial: Lamarck envisaged the Chain of Being in temporal as well as structural terms, but there was actually very little scope for time to penetrate the ontological identity of species, since change was largely determined by the action of an invariable force, the *tendance à la complexification*, which inheres in, and acts regularly upon, vital matter. More properly historical processes were only allowed a limited role, in the form of environmental factors that cause local bunchings on a chain which would otherwise be perfectly regular. Hence François Jacob's description of Lamarck's system as existing in an 'état stable dynamique'; a state of dynamic equilibrium.[2] Lamarck's temporalisation of the domain of living things gave rise, we might say, to an essentially scientistic rather than a properly historical conception of organic becoming for the animal world. Darwin's model, by contrast, did just the reverse.

---

[1] Richard Somerset, 'Transformism, Evolution and Romanticism', *Nineteenth-Century French Studies* (2000–2001), 29, 1–20.
[2] François Jacob, *La logique du vivant: une histoire de l'hérédité*, Paris, 1970, 168.

Meanwhile, French historiography – I went on to argue – had also evolved in response to the organicist outlook. Organicism manifested itself here in the new importance given to *mass* phenomena – the Nation, the People, the Race – which could now be conceived metaphorically as having proper organic existences, that develop spontaneously and independently over time. It was a tactic which might allow historiography to develop scientific pretensions since this type of growth or change might be held to depend entirely upon natural law. Theoretically, it was possible to move towards an organicist history of a People or of a Nation which effectively banishes historicity from the ontological formation of these entities, just as Lamarck's theory effectively did for animals. But in fact, it is much harder for a historian to arrive at such a result, simply because his object of study cannot quite escape the human or artificial basis to its ontology: Nations and Peoples are things that are created in large part by a series of more or less random local actions, and so it is difficult to submit their ontological formation to scientistic or linear modes of becoming. Because of this structural circumstance, it happens – more or less accidentally – that certain organicist histories end up promulgating a version of becoming that is conceptually quite similar to the Darwinian.

That is roughly what I argued in my earlier article; what I aim to do here is to integrate into this analysis the parallel developments characteristic of British science and historiography, and to see how they compare to the patterns I have identified for France. I will show that in Britain just the opposite tendency prevailed: that properly historical becoming was developed by Darwin in the natural sciences, while historical treatments of 'organic becoming' tended, ironically enough, to be more scientistic. I will try to show why this should be so.

I will proceed by comparing the work of Jules Michelet and Thomas Carlyle, two important historians, one French and one British, who were active in this period, and whose writings were influential amongst their contemporaries. These two figures provide an interesting comparison since their work presents many points in common. They both wrote histories of the French Revolution, for example, and their approaches were similar enough for Carlyle's French translator to claim that 'Carlyle is a Michelet to the fourth power, and Michelet is an exclusively French version of Carlyle'.[3] It is probably safe to say that, as historians, they were chiefly similar to one another in those respects which might be considered to qualify them as 'Romantic' figures – and in fact each of them has routinely been reckoned the *chef de file* of Romantic historiographers, each in his own country. Of course the label of 'Romantic' is a complex one that means many things to many people, but we can at least identify as an important point for our purposes an explicit application of the organicist outlook in

---

[3] *Histoire de la Révolution Française*, trans. Jules Roche, Paris, 1912, xvi. All translations from the French are my own.

their conceptions of the historical process of becoming. However, there are important underlying differences in their historiographical modes, and these differences can act as a kind of a gauge of the more general tendencies in innovative historiographical method in these two countries during the pre-Darwinian period. Of course there were plenty of other historians doing other things in both Britain and France, and it cannot be claimed that these examples provide anything like an exhaustive picture of the whole range of historiographical discourses of the day. I have chosen these two as the classic British and French organicist historians.

I will concentrate in particular on two short theoretical pieces both published in 1831: Michelet's *Introduction à l'histoire universelle* and an essay by Carlyle entitled 'Characteristics'. In this comparison I will show how the two historians had a similarly organicist outlook but nevertheless constructed their modes of becoming in radically opposed ways. At the heart of this difference is a strongly contrasted understanding, or location, of human free will; and this distinction will be particularly useful for my analysis.

I will start with Michelet, since he provides a clear theoretical statement of how organicism works in history. In 1869, looking back on the 17-volume *History of France* which he had started writing in the aftermath of the July revolution of 1830, Michelet describes the initial conception of the project in the following terms:

> In those memorable days, there was a great light, and I perceived France.
>
> She had annals, but no history. ... Nobody had yet beheld the unity of the natural and geographical elements that constitute her. I, first, saw her as a soul and as a person. ... Life depends on one all-important and demanding condition. Life is not life unless it is complete. Its organs can only work if they work together. Our vital functions are connected; the one supposes the other. If just one is absent, there is no longer any life. People used to believe that life might be isolated by the scalpel, that each part of our systems could be studied on its own. This is not in fact possible, for everything influences everything else.[4]

These are the familiar terms of the new physiology and comparative anatomy that had been developed by Bichat, Lamarck and Cuvier, with its characteristic insistence upon the indivisibility of the organic system. The Nation can be considered diachronically as an emergent being, or synchronically as a complex being at a given stage of development. Either way, the 'organs' which make up the Nation-as-an-organism, are the various racial components that have combined to form that People. Hence the importance of the *Tableau de la France*, the 'moral geography', as Michelet later called it, that occupies the whole of the third volume of his *Histoire de France*, and

[4] Jules Michelet, 'Préface de 1869' for *Histoire de France*, in *Oeuvres Complètes*, 22 vols, Paris, 1971–82, iv, 11.

which takes the reader on an imaginary tour of the provinces, showing how each one's distinctive racial and temperamental particularities contribute to the Nation's character as a whole. It is interesting to note that because the Nation is an organism and not a curious stone or crystal, the whole is something more than – or something other than – the sum of the constituent parts. In volume four, after the *Tableau de la France*, he provides the following summary:

> Race upon race, people upon people; Galls, Kymrys, Bolgs; in one place Iberians; in another place Greeks and Romans; and then finally the Germanic tribes. Having said all this, has one said France? Almost everything remains to be said. France made herself out of these elements, from which any other combination might have resulted. The same chemical elements compose oil and sugar. The elements being given, the whole is not yet given; there remains the mystery of the particular form of existence. How much more must we take these considerations into account when we are dealing with a living and active mixture, such as a Nation; with a mixture capable of working on itself, of modifying itself? This work, these successive modifications, by which our nation advances by spontaneous transformation, is the proper subject of a history of France.[5]

So this organism, this living, breathing France is not the accumulated result of a series of brute facts or events piled on top of one another: France is the being that, in a sense, digests those events, and makes its own body of them, in its own more or less unpredictable way.

This capacity for assimilation or digestion is in a sense the natural norm that applies to all Nations; but it also turns out – moving now to the *Introduction à l'histoire universelle* – that France is the archetypal Nation in this respect. The particular 'genius' of France is its capacity for 'universal receptiveness'. Assimilation is the most fundamental organic activity, and France is the most active of Nations; it provides a kind of organic *résumé* of Europe, and so of progressive humanity. It is the crucible in which emergent patterns form.

> The sign and the guarantee of a living organism – the power of assimilation – is here present in its highest degree: the France of the French has managed to attract, absorb and combine the Frances of the English, the Germans and the Spanish, that surrounded her. She has neutralized the one by the other, and has converted them all into her own substance. … Her particular genius is a capacity for [organic] action, and that is why the world belongs to her. … France acts and reasons, rules and struggles. She moves the world. She makes history, and her history tells the story of humanity.[6]

---

[5] Michelet (1971–82), iv, 182.

[6] *Ibid.*, *Introduction à l'histoire universelle*, ii, 247–9.

Of course all this is a rather disingenuous way of making the quality or value which Michelet believes to be the end of human history appear to be a natural value that inheres in the spontaneously arising nature of things, and not merely a value of his own construction. In this respect, his history is essentially 'whiggish'[7] in character, even if that term sounds odd in a French context. For Michelet, the proper end of history is the establishment of liberty as the defining human value; and of course the French Revolution is interpreted as the critical event that made the proper condition of humans realisable. Because liberty is seen as an absolute value (its value assured by reason), Michelet has no qualms about using it retrospectively as a yardstick by which to assess former ages. Hence his difficult relationship with the Middle Ages, for which he has a certain paternalistic sympathy as the historian responsible – as he saw it – for its organic revivification, but which he also detested as the age that had 'hunted Liberty down in the name of Satan'.[8] Hence also his admiration for Joan of Arc, which derives from his seeing in her a kind of failed premonition of the popular movements of 1789–93. In the chapters of the *Histoire de France* devoted to her story, he calls her 'the heroic Evangelist of the people, the living prophecy of the Revolution'.[9]

But despite this unabashed whiggishness, Michelet's historiographical system remains an essentially unguided one – for the simple reason that his politics requires that it must be so. The choice of liberty as his fundamental human value constrains Michelet in terms of the kinds of historical dynamics he can envisage. If the proper element of humans is liberty, he must be able to exercise free will; he must be able to construct his own environment. And if this is so, his nature – as a human being – is radically distinct from the norms of natural nature. The *Introduction à l'histoire universelle* starts with an explicit statement of this attitude, and it even goes so far as to place human nature and natural nature in mutual conflict:

> With the world began a war which will end with the world, and not before. It is the war of Man against Nature, of spirit against matter, of liberty against fatalism. History is nothing else than the story of this interminable struggle. … That which should encourage us in this struggle without end is the fact that in the end the advantage lies with us. Of the two adversaries, one never changes, while the other does change and becomes stronger. Nature stays the same, while every day Man extends his advantage over her.[10]

The human beings's liberty is to be contrasted to nature's fatalism. Nature is material and determined; the human being is rational and free. Michelet is not afraid of the consequence that freedom therefore seems *artificial.* He states that:

---

[7] I use the term in the sense elaborated by Herbert Butterfield in *The Whig Interpretation of History*, Harmondsworth, 1973.

[8] Michelet (1971–82), 'L'héroisme de l'esprit', iv, 34.

[9] *Ibid., Histoire de France*, vii, 76.

[10] *Ibid.*, ii, 229.

> The most artificial, which is to say the least fatalist, the most human and free part of the world, is Europe; and the most European of the European nations is my country, France.[11]

Man's capacity to construct himself, a capacity that derives from free will, is a heroic god-like capacity that sets him radically apart from the trammelled ways of brute nature. Hence his definition of what he calls 'the heroic principle' as the process of:

> humanity making itself. The living force [of liberty] which is Man, creates itself by its actions, by its works, by its cities and its gods (which are also its actions.)[12]

Michelet is whiggish, then, to the extent that his history is built upon the presupposition that human history is or should be the history of the attainment of political liberty, and that the course of history is only correctly understood when interpreted in these terms. But because of his primordial belief in individual freedom as the essential *political* reality, he is obliged to maintain an active role for unpredictable human action in the unfolding of human history – and even to keep out natural law as a type of process that encroaches unacceptably upon that essential freedom.

Thomas Carlyle's approach to history was just as organicist as Michelet's. He too conceived of humanity as having a developmental trajectory through time, similar to that of the growth of an individual living thing. In 'Characteristics', for example, he writes that society is:

> the vital articulation of many individuals into a new collective individual ... To figure Society as endowed with life is scarcely a metaphor; but rather the statement of a fact by such imperfect methods as language affords.[13]

On the other hand, Carlyle clearly differs from Michelet in his political and moral reaction to the French Revolution. For Carlyle, as for most of the reading public in Britain, the revolution did not signify the ultimate liberation of humankind, but stood instead as a terrible demonstration of the consequences of political and moral degeneration. However, Carlyle was no reactionary: the evils of the revolution were not properly attributable to the mob, nor even to the sceptical materialist philosophy of the eighteenth century, but rather to the essential falseness of the preceding moral order of things. 'The Encyclopedists did not produce the troubles of France', he

---

[11] *Ibid.*, ii, 247.

[12] *Ibid.*, 'L'héroisme de l'esprit', iv, 41.

[13] Thomas Carlyle, *Critical and Miscellaneous Essays*, 7 vols, London, 1869, iv, 11. Carlyle's organicist outlook is perhaps more famously exemplified by his usage of the Nordic myth of 'the tree Igdrasil' in *On Heroes* (1841) and in *Past & Present* (1844).

writes, 'but the troubles of France produced the Encyclopedists, and much else'.[14] The dangerous philosophy of the eighteenth-century radicals and the rebellions they provoked were the natural consequences of a rotten system which had alienated humans from the true relationship to others, to God and to nature. The revolution was a just punishment for these abuses; it would bring France back to a proper balance, and would serve as a warning to the rest of the world of the consequences of ignoring the dictates of the immediate moral sentiments, which are the human being's surest insight into the eternal nature of things. The final message of Carlyle's *The French Revolution* (1837) was that we must 'go and do *otherwise*'[15] if similar episodes are not to be repeated.

Like many other British commentators in that revolutionary and industrialising age, Carlyle put German idealism to the service of an essentially moral message; a message which turned around a reaction against mechanism. Humankind had gone wrong because it had allowed everything about its existence to become mechanical – not just its modes of production, but also its modes of thought, and even its modes of feeling.[16] The *philosophes* had been responsible for producing this tendency by reducing the causality of the phenomenal world to material considerations, and making humans the independent masters over it by the rational exercise of free will. This development marked, for Michelet, humankind's passage into an adult or perfected condition; while, for Carlyle, the same change signalled its ruin. To make human beings' will the master of the universe is to misunderstand the nature of things: it is to disengage the human being from the world; to render them incapable – individually or collectively – of rightly perceiving what they *are* in the grand scheme of things.

So where does this leave free will in Carlyle's world? As a historian of the French Revolution, and fascinated by the unpredictable world-changing chaos sparked in 1789 by apparently minor local causes, he was certainly not inclined to deny humankind all purchase upon the course of history. But how then is he to reconcile these observed deviations with his moral conviction of the fundamental connectedness of the ideal and the actual orders of things? In 'Characteristics', Carlyle attempts to deal with this characteristic problem of idealist history (i.e., how to deal with regressive periods) by applying a form of the Saint-Simonian theory that the course of history resembles a sine wave, with alternating constructive and critical periods. For Carlyle, the positive

---

[14] Carlyle (1869), iv, 18.

[15] Thomas Carlyle, *The French Revolution*, 2 vols, London, 1906, ii, 382.

[16] 'Not the external and the physical alone is now managed by machinery, but the internal and spiritual also ... Men are grown mechanical in head and in heart, as well as in hand. ... Their whole efforts, attachments, opinions turn on mechanisms, and are of a mechanical character. ... This is not a Religious age. ... Our true Deity is Mechanism. It has subdued external Nature for us, and we think it will do all other things.' 'Signs of the Times' (1829) in Carlyle (1869), ii, 234–6, 245.

periods are *dynamic* and the negative ones *material*.[17] By 'dynamic' he means that their spirit flows directly from nature. These are heroic, intuitive ages in which humans *unconsciously* follow the path intended for them by the world soul (or by nature). It is the instinctive following of this path that constitutes true progress, since humans thus close to nature can be properly creative; can gain inspired insights into the eternal order of things, and so advance humankind towards its proper consummation in the eternal ideal. The negative periods are 'mechanical' precisely because they lack this naturalness. They are ages of conscious doubt and questioning, whose hesitant path derives from logical or philosophical speculation; that path is artificially constructed, and does not develop organically. Human beings' self-conscious condition here limits their existence; they are too conscious of their beings to *be*. These periods cannot be progressive. Cut off from nature by their very self-consciousness, the human being who is led by logic rather than instinct cannot *create*, they can only *manufacture*. They are not free. In brief: for Carlyle, it is only when the human being is in contact with nature, only when they are appropriately *restricted* by the spirit of nature, that they can be free.

So, comparing the two, we see that Michelet's free will struggles against nature, and Carlyle's cannot exist outside it. While Michelet's prioritisation of the political notion of individual liberty pushes his organicist history into a position in which the human being is placed in opposition to nature as an active player in the dynamic of becoming, Carlyle's moral imperative to block materialism pushes his equally organicist history into the opposite position in which the overall sense or direction of the historical dynamic is controlled from outside the system, by a higher or other intelligence. Michelet must put history into the hands of the human being if their liberty is to have any meaning, and Carlyle must take history out of the hands of the human being if their morally-conceived normative curve is to have any credibility as an eternal given. Counter-intuitively, perhaps, it is thus Carlyle's morally oriented history rather than Michelet's more materialistic political history which is drawn towards the linear or scientistic way of dealing with organic development.

It is interesting to notice that the logic of Carlyle's system is remarkably similar to Lamarck's. Carlyle saw human history as the product of two interacting and potentially contrary forces; the smooth progressive tendency of the underlying Ideal force of things, interacting with the locally distorting results of individual human action. Lamarck's transmutatory model also worked on two such counteractive forces: the *tendance à la complexification* which would theoretically produce a perfectly smooth

[17] The distinction between the 'dynamic' and the 'material' were fundamental to the organicist thought of this period. Originating with the new life sciences, the distinction had already been borrowed by idealist thinkers such as Goethe and Coleridge.

Chain of Being, if it were not countered by the opposing local effects of environmental considerations on individuals and small groups, which tends to produce localised bunching on the Chain. Lamarck's theme that essential continuity is visible in the masses and not in the individuals finds a constant echo in Carlyle's distinction between the long-term or normative course of history and the short-term actual course of things. (Of course there is also an essential difference in that Carlyle's ideal history finds its guidedness in a transcendental source, whereas Lamarck's guidedness is immanent in its source and material in its operation. In this respect, Carlyle's model resembles Chambers' rather than Lamarck's.[18])

On the other hand, Michelet's a priori desire or need to talk up the free will of the human being, so that liberty as a goal for humanity can make sense, is compelled to resist any prior guidedness, whether it be material or transcendental in origin. And it is this quality which makes his mode of organic becoming resemble that which was later envisaged by Darwin.

Although the claim might seem paradoxical, perhaps the clinching argument in this affair is that neither Carlyle nor Michelet finally responded to scientific developmentalist theories in the ways that my analysis might seem to suggest. (Both men, by the way, lived long enough to know about the Darwinian hypothesis.) Carlyle, despite the a priori resemblance of his developmental outlook to Lamarck's, remained unreceptive to theories of transmutation in any form; and Michelet, despite the a priori resemblance of his developmental outlook to Darwin's, became deeply, even mystically involved in a distinctly Lamarckian-flavoured universal evolutionary picture.[19] But of course we must bear in mind the distinction between the worlds of human and natural affairs; the important distinction that I made from the start. Carlyle and Michelet think one way about human affairs and quite another about animal or natural affairs. If Carlyle was unreceptive to transmutatory theories – even to one like Chambers', which potentially lent the basic Lamarckian model the requisite moral flavour – this was, I think, because he was basically not interested in the world of natural creation. As a moralist he was primarily interested in the world of thought. What is fascinating about the human being as a moral being – i.e. a thinking, changing being – is their changefulness; their existence in between absolute natural determined-ness and absolute human freedom. The merely natural world, which remains unthinking and unchanging, is simply not interesting to Carlyle; and the theories of a Chambers or a Darwin simply do not succeed in bestowing on this domain an interesting sort of changefulness, since there is not and cannot be any dimension of moral choice

[18] Robert Chambers, *Vestiges of the Natural History of Creation and Other Evolutionary Writings* (ed. James A. Secord), Chicago, 1994. Chambers' *Vestiges* was published anonymously in 1844.

[19] See his late series of 'Natural Histories': *L'Oiseau* (1856), *L'Insecte* (1857), *La Mer* (1861), *La Montagne* (1868).

involved. Michelet, on the other hand, in switching from the human to the natural world, was able fully to indulge the impulse towards guidedness that seemed to be an irresistible corollary of systematic material organicism; an impulse which he was prevented from indulging in his histories by his political a prioris about the nature of the human being.

Of course I cannot claim that there is any kind of direct influence at work here; I am certainly not claiming that Michelet or any other historians like him had any direct influence on Darwin. I am dealing with patterns in the modes of conception of organic becoming, not with intellectual influence. But I can perhaps see in this analytical picture something about why the Darwinian approach to organicist becoming as applied to the natural world was better suited to the British context than to the French. In general terms, I might say that in the British context, Darwin's strategy of basically ignoring the questions of ultimate origin and guidedness was, first, *desirable*, morally speaking; and, second, *possible*, methodologically speaking. It was desirable to avoid these issues because they were particularly apt to raise the spectre of materialism and all its moral implications.[20] At the same time, the intense interest shown by non-scientific 'specialists' in moral questions such as Carlyle helped to create the option of non-engagement as a legitimate methodology for scientists. Not only could Darwin hope to avoid getting in trouble by steering clear of those questions, he could also hope that his choice would be taken as a token of respect for the separate authority of those who had chosen to make the moral dimension of 'organic becoming' their special concern.

In France, by contrast, the rather less violent reaction against materialism left everyone, scientists and historians alike, in a rather different situation. Historians like Michelet, as I have shown, could remain attached to a fundamentally materialist conception of humanity – or, at least, they did not necessarily feel the need expressly to combat materialism by invoking a guiding spiritual dimension to human affairs. This greater laxity provided the possibility that organicist history could be formulated in a less directed manner. In science, however, the same relative openness towards materialism had precisely the opposite effect. The business of science remained the provision of a material explanation of the phenomenon in question. Hence the primary goal of Lamarck and subsequent French naturalists such as Etienne Geoffroy Saint-Hilaire who advocated that the mutability of species was not merely the elaboration of a theory explaining the mechanism of transmutation, but, much more ambitiously, the provision of a material explanation of 'life' as a natural phenomenon. The limited Darwinian approach was unlikely to be adopted by a French naturalist because there was no compelling moral need, nor any methodological precedent for such a tactic.

---

[20] Chambers' attempt to construct a non-materialist developmental model had been seen by many as little more than the same old materialism in disguise.

The scientist was not put under moral pressure to avoid the issues of origin and direction; if anything, he found himself under methodological pressure *not* to go this way. The operation of this epistemologically more demanding standard of scientific explanation is plain in the familiar rejection of Darwin's theory as *unscientific* by the French establishment. *The Origin* was criticised by one prominent academician, for example, in the following terms:

> Mr. Darwin has just published a book on the Origin of Species. The illustrious author thinks that the natural species is flexible. Unfortunately he does not tell us what he means by species, nor does he suggest any definite criterion by which to define the term. ... Mr. Darwin writes a book on the origin of species, and what is missing from this book is precisely the origin of species.[21]

It is interesting to note that this view has not entirely disappeared today. André Pichot, a modern commentator, persists in regarding the epistemological relaxation represented by Darwin's approach as compared to Lamarck's as a *loss*. He says:

> Curiously, the theory of evolution is still typically presented in these [Darwinian] terms. ... The epistemological need for an evolutionary theory capable of bringing modern complex beings back into the domain of physical law has been completely forgotten.[22]

In Britain, then, properly historical modes of conceiving organic becoming were more likely to occur in the sciences than in history; and this for moral reasons. In France, on the other hand, the opposite was true – such modes were more likely in history than in science; and this because the absence of that moral imperative meant that history could remain historical and science scientific.

---

[21] Pierre Flourens, 'De l'origine des espèces, ou des lois du progrès chex les êtres organisés, par Ch. Darwin', *Journal des Savants*, October 1863, 622 and November 1863, 703.
[22] Editor's introduction to Lamarck, *Philosophie zoologique*, Paris, 1994, 45.

Chapter 10

Charles Darwin: A Christian Undermining Christianity?

Momme von Sydow[1]

The relationship between science and belief in Charles Darwin's thought is much more intricate than one might assume solely on the basis of a general conflict of supposedly opposing ideas. In this article, I would like to give an account which makes the interplay of ideas – both positive and negative – between scientific and religious thought in Darwin's philosophy intelligible, by providing evidence that there is a self-undermining dynamics of ideas between belief and science. The starting point for my discussion will be to consider how Darwin's particular theory of evolution was inspired by the tenets of a secularised standpoint and even a manifestly religious background. I then consider how 'Darwinism' undermined its author's former Christian world-view, on which it was partly based. This leads me to raise the question of how the resulting undermining of Darwin's religious views, in turn, contributed to (or perhaps allowed) modifications to his biological theory. Finally, the dynamics of self-undermining ideas between belief and science will be briefly considered in the context of a general understanding of the history of science.[2]

From Natural Theology to Natural Selection

Darwin's theory of evolution is itself not a *creatio ex nihilo*, but one of the greatest empirical, and also theoretical, syntheses known in the history of ideas. The theoretical aspects of this synthesis are not exclusively based on scientific ideas, but, despite the resulting anti-religious inclinations, on religious ones as well.

Charles Darwin was born into a wealthy family of intermarrying Darwins and Wedgwoods. This family had been influenced by Unitarianism, which can be characterised by a critical attitude towards the Trinity, but also a firm belief that God's

---

[1] I would like to thank David Knight, John H. Brooke and Trevor Levere for their advice and constructive criticism, and M.D. Eddy for many comments.

[2] The so-called 'Darwin industry' provides rich sources for researching every minutiae of Darwin's life and intellectual development.

benevolence is expressed in the material world.[3] Accordingly, the young Charles was more interested in the varieties of God's material creation than in the interpretation of the Gospels. Nonetheless, when he was preparing to study at Cambridge, where he intended to become ordained as an Anglican priest, Charles clearly (and perhaps naively) believed in God. After reading John Bird Sumner's *Evidences of Christianity*, he noted that there was 'no other way except by [Jesus'] divinity' of explaining the historical evidence provided by the Gospels.[4] Later, Darwin went on to write in his *Autobiography*, 'I did not then in the least doubt the strict and literal truth of every word in the bible'.[5]

Because of this religious background, several of the following sections will consider three of the main influences on Darwin's biological theory which had a direct or indirect religious origin: Paley's belief in the divine design of nature; the conviction that God rules by laws which are eternal, universal and unchangeable; finally, Malthus' principle of population, partly presented as a theodicy.[6]

---

[3] Charles Darwin's maternal grandfather, the pottery patriarch Josiah Wedgwood was a convinced Unitarian and member of the Lunar Society. He maintained close contact with the chemist and influential Unitarian Joseph Priestley (1733–1804). Hence, Josiah appointed a Unitarian minister for his own school. Charles Darwin's paternal grandfather Erasmus Darwin, also a dissenter, sent his son Robert (Charles' father) to this school. Also, Charles' mother, Susanna Wedgwood, was educated there. Although Charles himself intended to enter the Church of England, he was first sent to a day-school in Shrewsbury, run by the minister of a Unitarian chapel (N. Barlow, ed., *The Autobiography of Charles Darwin 1809–1882. With original omission restored*, New York, 1958 (written 1876), comment of F. Darwin, 22). Darwin's early Unitarian piety was encouraged by his sisters Catherine and Caroline (cf. their letter to Charles: F. Darwin, ed., *Life and Letters of Charles Darwin*, 1897, reprinted in *The Works of Charles Darwin*, vols 17 and 18, New York, 1972, vol. 1, 11th April 1826).

[4] A. Desmond and J. Moore, *Darwin*, London, 1992/1991, ch. 1, 8 f. and ch. 4, 48 f.

[5] Barlow, ed. (1876), 57. On board the HMS Beagle, Darwin still believed in the immutability of species and 'was quite orthodox'. He remembered being laughed at by some of the officers, 'for quoting the Bible as an unanswerable authority on some point of morality'. (*Ibid.*, 85).

[6] The significance of natural theology in establishing a framework for Darwin's biological theory was first stressed by W.F. Cannon in 'The Bases of Darwin's Achievement: a Revaluation', *Victorian Studies*, IV, 1961, 109–34, esp. 127–30. Subsequently, Paley's influence on Darwin has been acknowledged, in particular by: J.H. Brooke, 'The Relations Between Darwin's Science and his Religion', 40–75, in J. Durant, *Darwinism and Divinity*, Oxford, 1985; D. Ospovat, *The Development of Darwin's Theory; Natural History, Natural Theology and Natural Selection, 1838–1859*, Cambridge, 1995/1981; R. Young, 'Darwin's Metaphor: Does nature select?', *The Monist*, 1971/1985, sec. IV–VI. The account given here is similar to Brooke's account, since he has also stressed that there are both positive as well as negative interactions of scientific and religious ideas in Darwin's thought. Additionally, I intend to place these interactions in a temporal order of a dynamic of self-undermining ideas between belief and science.

Darwin's Early Belief in Divine Design and its Relation to his Later Pan-Adaptationism

The concept of perfect adaptation and divine design in nature, as developed particularly in William Paley's (1743–1805) *Natural Theology*, was the religious basis for Darwin's early pan-adaptationism. At Cambridge, Darwin occupied the same room at Christ College as had Paley.[7] Darwin was formally required to read Paley's *Evidences of Christianity* and *Principles of Moral and Political Philosophy* for his BA degree. He learned the *Evidences* by heart and answered all of the questions on Paley particularly well. Moreover, Darwin, who was basically not a very ambitious student, read Paley's *Natural Theology* voluntarily and with delight.[8] He also knew the *Natural Theology* almost by heart[9] and read it repeatedly, even after he had passed his exams. In the late 1820s and early 1830s, England's natural theology was still in bloom; in particular, in the natural sciences community at Cambridge, amongst which the young Darwin spent most of his time and where he became imbued with thoughts from Paley's *Natural Theology*.[10] Even later, Darwin wrote in a letter that he 'hardly ever admired a book more than Paley's *Natural Theology*', and that the careful study of Paley's works was the only part of his academic course at Cambridge which left a permanent impression on him.[11]

Paley held that there were two sources for collecting evidence for the existence and attributes of the Deity. Firstly, there were the Scriptures. These gave an account of the literal historical truth of Christian miracles and he treated this topic in his *Evidences of Christianity*. Secondly, there were the 'designs and dispositions from his [the Creator's] works; or, as we normally call it, the light of nature', which he treated in his *Natural Theology*.[12] Paley's *Natural Theology* was particularly important for Darwin, who also worked on the 'book of nature'. The argument in Paley's *Natural Theology* runs from perfect adaptation or design in nature to the existence of God as an omniscient, divine designer. Paley argued that the mechanical design of a watch found on a heath would testify to the existence of an artificer. By analogy, the perfect design found in nature, found for instance in the complex function of an eye, which Paley conceived

---

[7] A. Desmond and J. Moore (1992/1991), 63–4; cf. Darwin, ed. (1887), vol. 1, ch. IV, 139; F. Burkhard, et al., eds, *The Correspondence of Charles Darwin*, Cambridge, 1985, vol. 1, 70–71.

[8] Barlow, ed. (1876), 59; Burkhard, et al., eds (1985), vol. 1, 75; Burkhard, et al., eds (1985), vol. 1, 112.

[9] Burkhard, et al., eds (1985), vol. 7, Letter to J. Lubbock, 22nd November 1859, 388.

[10] Cf. A. Fyfe, 'The Reception of William Paley's "Natural Theology" at the University of Cambridge', *British Journal for the History of Science* (1997), esp. 321, 329 f., 335.

[11] Barlow, ed. (1876), 59.

[12] Cf. already W. Paley, *The Principles of Moral and Political Philosophy*, first: 1785, ch. IV, in *The Works of William Paley, D.D.*, Edinburgh, 1842.

along mechanical lines, proved the existence of an omniscient designer.[13] Also, Paley's proof of the benevolence of the Deity hinges on the idea of perfect adaptation. Since Paley understood adaptations at the individual level to be 'beneficial', and since he observed adaptations mainly at the individual level, he was able to conclude that, 'in a vast plurality of instances in which contrivance is perceived, the design of the contrivance is beneficial'.[14]

The young Darwin found the 'argument of design in nature, as given by Paley, … conclusive'.[15] Even when Darwin became convinced of the transformation of species in 1837 (like his grandfather Erasmus long before) and adopted his particular theory of evolution in 1838, he retained the Paleyan belief in universal adaptation.[16] In 1842, he suggested, in the first full sketch of his theory, that the secondary law of natural selection, which was 'impressed on matter by the Creator', was capable of 'creating individual organisms, each characterised by the most exquisite workmanship and widely-extended adaptations'.[17] Ospovat has argued that Darwin continued to believe in the Paleyan tenet that organisms are perfectly adapted after 1838 until at least 1844, and, only partly modified, at least until 1859.[18] In 1857, when Darwin confessed his theory to Asa Gray in a letter, he wrote (with obviously religious overtones) of 'a being' selecting for one end, during millions of generations and of 'an unerring power at work in Natural Selection …, which selects exclusively for the good of each' organism.[19] Also,

---

[13] W. Paley, *Natural Theology, or Evidences of the Existence and Attributes of the Deity Collected from the Appearances of Nature*, first: 1802, in *Works of William Paley* (1842). Particularly chs I, V, XXIII; on the attributes of the Deity see chs XXIII–XXVI.

[14] W. Paley, (1802), ch. XXVI, the first of the two propositions. Also W. Paley (1785), ch. V. Also Paley's individualism may have influenced Darwin.

[15] Barlow, ed. (1876), 87 and also footnotes 9 and 11.

[16] The dates of these turning points are largely accepted today: Cf. P.H. Barrett, et al., eds, *Charles Darwin's Notebooks, 1836–1844*, Cambridge, 1987; S. Herbert's introduction to Darwin's *Red Notebook*, 18; D. Kohn's introduction to *Notebook D*, 329–30 and note 28th September 1838, orig. 135e. Cf. also Barlow, ed. (1876), 83, 119–20. Sloan has correctly pointed out that Darwin's early concept of transmutation is reminiscent of Erasmus Darwin's, Schelling's, Oken's and Humboldt's concept of a creative nature. In contrast to Sloan, I would stress that Darwin nevertheless transformed this concept by integrating it into the respectable framework of British natural theology and of Newtonian science. P.R. Sloan. '"The Sense of Sublimity" – Darwin on Nature and Divinity', in J. Brooke, M.J. Osler and J.M. van der Meer, eds, *Science in Theistic Contexts. Cognitive Dimensions*, *Osirus*, vol. 16 (2001), 251–69.

[17] P.H. Barrett, et al., eds, *The Works of Charles Darwin*, London, 1986–89, vol. 10, *Sketch* (1842) 51, 52.

[18] D. Ospovat (1981/95), xv, ch. 3. Ospovat uses the term 'perfect adaptation' for absolute *and* relative adaptation. Moreover, he concedes that even perfect adaptation has its limits (73–4). This does not render Ospovat terminology inadequate, since even Paley concedes the existence of a few imperfections (1802, ch. V, XXVI).

[19] Burkhard, et al., eds (1985), vol. 6, 5th Sep. 1857, Appendix, sec. 3, see also sec. 2.

in the *Origin of Species*, a strong belief in universal adaptation (still with religious or moral reverberations) can be found: 'What limit can be put to this power, acting during long ages and rigidly scrutinising the whole constitution, structure, and habits of each creature, – favouring the good and rejecting the bad? I can see no limit to this power, in slowly and beautifully adapting each form to the most complex relations of life'.[20] Darwin's Paleyan belief in universal adaptation to the circumstances of life may also have encouraged him to principally attribute the causes of evolution to an extern source, to the environment of the organisms (as opposed to an inner, developmental force favoured by romantic biologists).[21] When constructing his theory of evolution, Darwin still hoped that this theory could at least be brought into harmony with a deistic belief in God. Correspondingly, in the resulting theory, organisms were still regarded as being machines that were almost perfectly designed and adapted; although the benevolent omniscient designer, God, had been replaced by the omnipotent process of natural selection.[22]

God's Eternal Law and Darwin's Process Monism

Another direct, or at least secularised, religious influence on Darwin was the belief in divine and preordained universal laws of nature. In the 1830s the alliance of religion and the sciences was still largely intact in England, but this alliance had acquired a deistic leaning. Paradoxically, the religious writings of Paley, in particular, gave a justification to Darwin's persistent search for a mechanical universal law of organic nature.

Paley, even in his early *Moral Philosophy*, had argued in favour of general rules and against a deification of accidents, such as bolts of lightening.[23] In a way, this continued the Christian agenda of demystifying the aspects of nature which had been formerly associated with pagan deities.[24] Even in the *Evidences of Christianity*, in

---

[20] M. Peckham, ed., *The Origin of Species: A Variorum Text*, Philadelphia, 1959 (orig. 1859, 1860, 1861, 1866, 1869, 1872), 1859, XIV, s. 99–100, also e.g. IV, s. 40.

[21] Cf. Paley's criticism of the 'absurdity of self-creation' (1802), ch. XXIII. For a discussion of the Darwinian (as well as Lamarckian) focus on external causes see: M. von Sydow, *Sociobiology, Universal Darwinism and Their Transcendence* (PhD thesis, University of Durham, 2001), 349.

[22] It has been pointed out that *some* uses of the term adaptation or survival of the fittest are tautological in stating the obvious truth of the survival of the survivor. Cf. St J. Gould, R.C. Lewontin, 'The Spandrels of San Marco and the Panglossian Paradigm', in *Proceedings of the Royal Society of London*, Series B, 205, 1979, 581–98; M. von Sydow (2001), sec. 9.1.

[23] W. Paley (1785), chs VII, VIII.

[24] The modern concept of nature ruled by universal laws can be understood as a mechanically transformed, Platonic-Christian concept of *machina mundi*, in which mechanical laws of nature have replaced the unchangeable Platonic forms, or ideas of God.

which Paley actually tried to prove the truth of Christian miracles in a literal, historical sense, he rigorously argued *against* the existence of any other supposed miracles.[25] Moreover, Paley, in *Natural Theology*, advocated the lawfulness of a mechanical world on theological grounds: a 'law presupposes an agent'. To Paley, the Book of Nature demonstrated that the Deity was acting according to general laws. Referring especially to Newtonian physics, Paley emphasised that these general laws were immutable, and if 'a particular purpose is to be effected, it is not by making a new law, ... but it is ... by the interposition of an apparatus, corresponding with these laws ...'.[26] According to Paley, knowledge of the attributes of the Deity (like omnipresence, eternity and unity) also rested on empirical evidence of the laws of nature (i.e. their generality, their inability to be changed and uniformity). Paley argued that evidence showed that 'the laws of nature every where prevail', that these laws 'are uniform and universal'. Therefore, Paley averred that these laws refer to an agent with corresponding attributes – an omnipresent and eternal God. As will become apparent in the next section, even Paley's defence of the benevolence of God (in spite of the existence of evil) is linked to the concept of God acting in nature by secondary laws.[27]

Likewise, the creed of preordained, unchangeable laws in nature was preached to Darwin in a rather secularised way. Already in the eighteenth century, the universal laws of motion, as discovered by Newton (a sage for Darwin's community of natural scientists) had become generally paradigmatic for a sober, deistic belief in God who acts by general and uniform secondary laws. This was the case, even though Newton himself had tried to maintain belief in an omnipotent God who intervenes directly.[28] The concept of simple, uniform and inviolable laws of nature was preached to Darwin in a partly secularised way, especially by Lyell and Herschel. Even before Darwin adopted his *particular* theory of evolution, he was in search of a universal and fixed law of nature which was analogous to Newton's universal laws of motion. He wrote in his notebook:

> Astronomers might formerly have said that God ordered, each planet to move in its particular destiny. – In same manner God orders each animal created with certain form

---

[25] W. Paley, *A View of the Evidences of Christianity*, first: 1794, in *Works of William Paley* (1842), proof of proposition 2 of Part I.

[26] W. Paley (1802), ch. I, sec. VII, ch. XXIII, similar in ch. XXV, ch. III, cf. ch. XXIII.

[27] *Ibid.*, ch. XXIV, also ch. XXV, cf. also ch. XXVI.

[28] Newton opposed Descartes' materialism and even argued that God continuously adjusts the planets preventing their gravitational collapse. In a *Scholium Generale* to the second edition of his *Philosophiae Naturalis Principia Mathematica* (1713) Newton argued that matter is passive and cannot produce gravitation – gravitation is to be attributed to God, the pantocrator (N. Guicciardini, *Newton*, Heidelberg, 1998). Newton, nevertheless, turned against the Aristotelian *causa finalis* and *causa formalis*.

> in [a] certain country, but how much more simple, & sublime power let attraction act according to certain laws such are inevitable consequen[ce] let animal be created, then by the fixed laws of generation, such will be their successors. –'[29]

Based on this statement and others like it from Darwin's writings, John C. Greene was right to call Darwin an 'evolutionary deist'. Darwin turned against any alternative evolutionary theory based on the concept of a directly intervening God or a *creatio continua* with almost a religious zeal. According to Darwin, the assumption that God would be concerned directly with the 'long succession of vile Molluscous animals' is 'beneath the dignity of him'.[30] After reading Malthus' *Principle of Population*, Darwin thought that this principle provided him with the universal law of evolution which he was searching for and he scribbled in his notebook that, 'since the world began, the causes of population & depopulation have been probably as constant as any of the laws of nature with which we are acquainted'.[31] In his *Sketch* (1842) and his *Essay* (1844), Darwin still continued to argue that the law he had found 'exalts' our notion of an omniscient creator and, even in the *Origin of Species*, his support of a universal law is partly justified in a religious way. Like William Whewell's *Bridgewater Treaties* and Joseph Butler's *Analogy of Religion*, the epigraph of the *Origin* claimed that, in nature, divine power becomes apparent in the existence of general laws.[32] Although Darwin, like others before him, broke with the static world view and proposed a theory of evolution, his theory of natural selection itself remained static, in proposing an unchangeable, almost preordained, evolutionary mechanism.[33]

The Influence of Malthus' and Paley's Theodicy

An additional religiously underpinned influence on Darwin was Thomas Malthus' principle of population. This struck Darwin as being an explanation of evolution and adaptation. Malthus' iron principle made a deep impression on Darwin, not only because it provided a mechanistic law appropriate to Newtonian science, differing from the Lamarckian explanation of adaptation and an evolutionary

---

[29] Barrett, et al., eds (1987), *Notebook B*, note from 1837, orig., 101. D.J. Depew and B.H. Weber in *Darwinism Evolving*, Cambridge, MA, 1995, have shown that Darwin's concept of natural selection is modelled along Newtonian lines, in an even more profound sense.
[30] Barrett, et al., eds (1987), *Notebook D*, 16th Aug. 1838, orig. 37.
[31] *Ibid.*, *Notebook E*, presumably 2nd Oct. 1838, orig., 3; dated according to D. Kohn.
[32] Barrett, et al., eds (1986–89), vol. 10, end of the *Sketch* and *Essay*. Peckham, ed. (1959), s. 1, 1.1b, cf. e.g. XIV, s. 259.
[33] For alternative proposals see, e.g. M. von Sydow (2001).

inner *Bildungstrieb*, but also because Thomas Malthus presented his principle as part of a theodicy, similar to the one Darwin knew in detail from Paley's *Natural Theology*.

It is generally acknowledged that Darwin got the final inspiration for the first formulation of his specific theory of evolution directly from re-reading Malthus' *An Essay on the Principle of Population* on 28th September 1838.[34] Malthus' principle of population says that human populations always increase much faster than their food supply. Malthus, also a political economist, had used this as an argument against the perfectibility of society, the aspirations of the French Revolution and the utopianism of W. Godwin and M. de Condorcet. In explaining evolution by means of scarce resources, a struggle for life and the survival of the fittest, Darwin applied Malthus' principle of population 'with manifold force to the whole animal and vegetable kingdom'.[35] Malthus, especially in the first edition of *An Essay on the Principle of Population*, but also later on, presented his critique of the utopians as a worldly theodicy.[36] He emphasised the positive effects of this harsh principle which 'prevents the vices of mankind, or the accidents of nature, the partial evils arising from general laws, from obstructing the high purpose of the creation'. Although the principle of population, according to Malthus, generally had positive effects, 'it is impossible that this law can operate, and produce the effects apparently intended by the supreme Being, without occasioning partial evil'.[37] Malthus' theodicy was based on the argument that God acted by universal secondary laws and by the principle of population. The resulting harsh conditions for the poor generally had positive effects, in that they checked the growth of population and imposed moral constraints. Thereby, Malthus had reconciled the existence of evil (and even the claim of an *unchangeable* existence of evil uttered in the first edition of his *Essay*) with that of a benevolent God.

---

[34] See Darwin himself: Barlow, ed. (1876), 120; also, e.g. his *Variation under Domestication*, 1875, Barrett, et al., eds (1986–89), orig., 8. Darwin's notebooks provide evidence of the date when he adopted a Malthusian approach: Barrett, et al., eds (1987), *Notebook D*, orig., 134 f., cf. 678. See also E. Mayr, *The Growth of Biological Thought*, Cambridge, MA., 1982, 477 f., but cf. 491 f.

[35] Peckham, ed. (1959), ch. III, s. 36.

[36] T. Malthus, *An Essay on the Principle of Population, as it affects the Future Improvement of Society with Remarks on the Speculations of Mr. Godwin, M. Condorcet, and other Writers*, 1st edn, 1798, 6th edn, 1826, in E.A. Wigley, et al., eds, *The Works of Thomas Malthus*, 8 vols, London, 1986. The presentation as theodicy is particularly striking in the widely known 1st edn of 1798. The chapter on natural theology has been dropped in later editions and its contents distributed to other parts of the essay (cf. D. Ospovat 1981/95, 66). Although Darwin in September re-read the 6th edn, he definitely was acquainted with the common interpretation of Malthus' *Essay* as theodicy.

[37] T. Malthus (1798, 1st edn), chs XVIII, XIX, 365.

The theodicy of Paley's *Natural Theology* is indeed similar to Malthus' theodicy. Paley, when treating the 'goodness of the Deity', conceded (despite his emphasis on perfection and happiness) that pain, privation and chance exist in numerous instances. Paley's theodicy is also based on the concept of general laws or rules: 'Of the Origin of Evil, no universal solution has been discovered ... The most comprehensive is that which arises from the consideration of general rules'.[38] The existence of a benevolent God could only be vindicated if he is acting by general laws, which may lead to partial evil, but whose effects are predominantly good. Moreover, the Malthusian topic of superfecundity also is essential to Paley's theodicy. For Paley the 'system of natural hostilities', e.g. animals preying on one another, is to be understood 'in strict connection with another property of animal nature, *superfecundity*'. This wastefulness may be justified as being also advantageous. Superfecundity, for Paley, firstly, 'tends to keep the world always full; whilst, secondly, it allows the proportion between the several species of animals to be differently modified, as different purposes require, or as different situations may afford for them room and food. ... Farther; by virtue of this same super-fecundity, what we term destruction, becomes almost instantly the parent of life'. Also, in relation to humankind, Paley explicitly took position similar to that of Malthus. For Paley, the harsh conditions of the poor did not constitute arguments against God, rather they showed that the world is in a 'state of probation', 'calculated for the production, exercise, and improvement of moral qualities, with a view to a future state'. Furthermore, Paley repeatedly stressed that there may have been some further consequences of the 'system of natural hostilities' hidden from us; and because of the benevolence which pervades on his account the general design of nature, we ought to presume that 'these consequences, if they would enter into our calculation, would turn the balance on the favourable side'.

From here it was not very far to Darwin's idea that the evolution of higher organisms – which he already took for granted – may justify superfecundity. When Darwin transformed static Malthusianism into a general evolutionary theory, he had found the very component which Paley had not; namely, a proper 'final cause' of superfecundity and struggle in nature in which the final cause 'must be to sort out proper structure & adapt it to change'.[39] Although Darwin soon dropped notions like 'final cause', he maintained aspects of a worldly theodicy in later writings. In his *Sketch* (1842), Darwin wrote, 'From death, famine, rapine, and the concealed war of nature we can see that the highest good, which we can conceive, the creation of the higher animals has directly come'. In this passage, Darwin then praises an omniscient creator, acting by the secondary law of natural selection.[40] The argument that the positive consequences

---

[38] The Paley quotations in this paragraph are taken from W. Paley (1802), ch. XXVI.

[39] Barrett, et al., eds (1987), *Notebook D*, 28th Sep. 1838, orig., 135e.

[40] Barrett, et al., eds (1986–89), vol. 10, end of the *Sketch* (1842), 51–2 (similar to the *Essay*, 1844).

of natural selection outweigh the negative ones is still, to some extent, to be found in the *Origin of Species*.[41]

Nevertheless, some discontinuities between the theodicies of Paley and Malthus and the secularised theodicy of Darwin remain. As has been noted by Bowler, even Malthus, who justified and demanded harsh conditions for the poor, did not favour the differential elimination of the unfit. Malthus intended to encourage 'slothful mankind' to work.[42] Darwin's resulting theory of natural selection, however, focused even less on the betterment of unfit organisms and more on their elimination.[43] Despite this change, the term 'natural selection' in Darwin's theory still mirrors the imagery of God's 'invisible hand'.

Darwin's Biological Theory Undermines the Christian Foundation on which his Scientific Theory had Partly been Based

Although Darwin's biological theory of pan-adaptationism and of an unchangeable law of natural selection was based on Paley's natural theology, it nevertheless became a main cause for Darwin's loss of faith. Darwin successively lost his belief in divine revelation, his confidence in a Paleyan benign conception of nature and, finally, even a belief in deism. Instead, he came to advocate a world-view based on a remorseless struggle for life and one which offered no outstretched hand to the losers.

Darwin had involuntarily proposed a rather diabolic 'principle of conservation of blindness, cruelty and wastefulness' that was reminiscent of Malthus, 'To prevent the recurrence of misery, is, alas! Beyond the power of man'.[44] In a view which stressed the unchangeable blindness of evolution, he held that it was absurd to regard God as omnipotent and as possessing foresight and benevolence. At best, God appeared to be a 'blind watchmaker'.[45] In his search for certain, eternal and ubiquitous 'laws of harmony',[46] Darwin finally proposed the Law of Natural Selection. Harmony became metaphysically based on and explained by an unchangeably cruel and wasteful struggle for life. In this light, the Christian credo 'As it was in the beginning, it is now, and

---

[41] At the end of the *Origin of Species*, in Peckham, ed. (1959), the argument of the *Sketch* and *Essay* is repeated, ch. XIV, 269–70 and ch. III, s. 165.

[42] P.J. Bowler, 'Malthus, Darwin and the Concept of Struggle', in *Journal of the History of Ideas* (1976), **37**, 631–50; Bowler, *Charles Darwin*, Oxford, 1990, 82–4.

[43] This is, however, less clear with regard to Darwin's early *Sketch* (1842) and *Essay* (1844) of his theory, according to which the environment triggers variation.

[44] T. Malthus (1798), 98.

[45] R. Dawkins, *The Blind Watchmaker*, London, 1991/1986.

[46] C. Darwin: Barrett, et al., eds (1987), 16th Aug. 1838. *Notebook D* (ed. by D. Kohn, 1987), orig., 36.

ever shall be: world without end' (*Gloria*) did not sound as hopeful as it had sounded before.[47]

It is difficult to trace the actual development of Darwin's religious views, since he, unlike some of his companions, remained reticent on religion and his crisis of faith. It seems that he was unwilling to give up his belief and did not want to evoke more dismay than necessary in public or in his pious wife, Emma.[48] In one of his private notebooks he explicitly advised himself to 'avoid stating how far, I believe, in materialism'.[49] Therefore, the development of Darwin's religious views has to be reconstructed from his private notebooks, his letters and his restored, relatively frank, *Autobiography*, which was intended for family use only.[50]

Considering these sources, Darwin obviously had difficulties in coming to terms with the materialist and atheist tendencies of his own explanation of evolution. In an early notebook, he wrote: 'love of deity [is the] effect of organisation. oh you Materialist!'.[51] This early self-characterisation would turn out to be truer than he could have foreseen at the time. Bearing this in mind, I will now trace how Darwin first lost faith in divine revelation, then his belief in deism, and how he finally came to regard himself (at the very least) as an agnostic.

Darwin himself stated in his private autobiography that he had gradually come 'to disbelieve in Christianity as a divine revelation'.[52] Darwin's early 'Paleyite Anglicanism, steeped in Unitarian nonconformity'[53] had been almost naively empirical

---

47 This section draws on M. von Sydow (2001), 33–6, 182–4.

48 Barlow, ed. (1876), 86, but 87, cf. e.g. Burkhard, et al., eds (1985), vol. 8, Letter to A. Gray, 22nd May [1860], 224; Darwin, ed. (1897), vol. 1, ch. VIII, F. Darwin comments on p. 276, and letter to F.E. Abbot, 6th Sep. 1871, 277, letter to E. Aveling, 13th Oct. 1880 mentioned in F. Burkhard, et al., eds, *A Calendar of the Correspondence of Charles Darwin, 1821–1882*, Cambridge, 1994, 2nd edn, 12757 and quoted in R. Young (1971/85), 20–21, 251. Cf. Young's conclusion, 21.

49 Barrett, et al., eds (1987), *Notebook M*, orig., 57; Burkhard, et al., eds (1994), Letter to J.D. Hooker 8th Sep. 1868, 6342 (cf. also F. Darwin, et al., eds, *More Letters of Charles Darwin*). Darwin's public attitude that a strict distinction between science and faith is most appropriate, appears to be due to his view that nothing good could result from connecting these domains of thought (cf. Burkhard, et al., eds (1994), 3208, 11766, 12931, 12088).

50 See footnote 2. Darwin's *Autobiography* was originally named *Recollections of the Development of my Mind and Character.* After Charles Darwin's death Francis Darwin, Charles son, edited the *Autobiography*, but purged it from offending passages (Darwin, ed. (1897), vol. 1, ch. 1, 8). The omissions have been restored in an edition of Charles' grand-daughter Nora Barlow (Barlow, ed. (1876), 21–145).

51 Barrett, et al., eds (1987), *Notebook C*, 1838, orig., 166. Cf. also later, e.g. *Notebook M*, orig. 136; Barlow, ed. (1876), 93; Darwin, ed. (1897), ch. VIII, letter to Graham, 3rd July 1881.

52 Barlow, ed. (1876), 86. Similar: Burkhard, et al., eds (1985), vol. 9, letter to B.J. Sulivan, 24th May [1861], 138. Letter to N.A. v. Mengden, 5th June 1879, in Burkhard, et al., eds (1994), 12088.

53 J. Moore, 'Of Love and Death: Why Darwin "gave up Christianity"', 195–229 in J.R. Moore, ed., *History, Humanity and Evolution. Essays for John C. Greene*, Cambridge, 1989, 196.

in character and, correspondingly, his disbelief in the truth of the Bible seems to have been mainly the result of empirical considerations. In Paley's *Evidences of Christianity*, the main arguments for divine revelation were based on empirical evidence for the literal historical truth of Christian miracles. An empirical literal understanding of the Bible, as opposed to a more metaphysical or symbolic one, was particularly susceptible to historical biblical criticism. The theological approach of Paley, particularly in his *Natural Theology*, was a characteristic product of the empiricist tradition.[54] Moreover, Paley, both in the *Evidences* and in his *Natural Theology*, demonstrated that there was no substantial evidence in favour of any non-biblical miracle and that nature was generally governed by laws. Darwin, as a naturalist, aimed to support these claims by showing that the organic world was also governed by law. But, ironically, this ended up undermining his own belief in *any* miracles: 'the more we know of the fixed laws of nature the more incredible do miracles become'.[55]

Subsequently, Darwin came to the conclusion that the argument from design in nature, as given by Paley, also 'fails, now that the law of natural selection has been discovered'.[56] Design or adaptation of organisms was explained by the laws of evolution. At first, Darwin still hoped that Paley's argument from design would remain successful in regard to the *laws* of nature – which do not themselves evolve.[57] But, only for a brief period did Darwin's deism become 'a featherbed to catch a falling Christian'.[58] Darwin soon came to see that there 'seems to be no more design in variability of organic beings and in the action of natural selection, than in the course which the wind blows'. Darwin, apparently forced by his own theory, rejected his former belief that the 'existence of so-called natural laws implies purpose'.[59] If the world was essentially a 'struggle for existence' or a 'war of nature',[60] if harmony was universally based on conflict, then a creator – as understood on the basis of his creation – did not appear to be a merciful benevolent deity, but rather an evil demiurge or a demon, whose views rested upon misery.[61] Descartes' hypothetical deceitful demon appeared to be harmless in comparison. Darwin indeed castigated himself as a 'Devil's Chaplain', 'writing on the clumsy, wasteful, blundering law & horrible cruel works

---

[54] M.D. Eddy, 'The Rhetoric and Science of William Paley's Natural Theology', *Literature and Theology* (2004), **18**, 1–22.

[55] Barlow, ed. (1876), 86.

[56] Barlow, *ibid.*, 87. Cf. Darwin's disputes with Gray and Lyell on divine forethought, Burkhard, et al., eds (1985).

[57] Cf. Barlow, ed. (1876), 87–90, 92–3.

[58] Cf. A. Desmond and J. Moore (1991), ch. 1, 256; J. Moore (1989), 209, 216, 221.

[59] Barlow, ed. (1876), 87; Darwin, ed., ch. VIII, Letter to Graham, 3rd July 1881; cf. Barrett, et al., eds (1986–89), *Variation under Domestication* (1875), 426–8.

[60] Peckham, ed. (1959), heading of ch. III, ch. III, s. 145, ch. IV, 10, 393, ch. III, 165, ch. XIV, 296.

[61] Cf. e.g. R. Dawkins, *River out of Eden*, London, 1995/1988, 109.

of nature'.[62] Working through the consequences of his theory even gave Darwin migraines.[63] Darwin, when describing his loss of a benign view of nature, confessed that he also had become, so to speak, colour-blind even to grand scenes of natural beauty, which formerly had made an overwhelming impression on him.[64] Darwin's ideas on religion nevertheless fluctuated and, when writing the *Origin*, he still hoped that an argument similar to Paley's and Malthus' theodicy would remain viable. He thought that maybe God as *prima causa* could be justified if happiness decidedly prevailed over suffering.[65] But Darwin also became increasingly less convinced of this point: 'I cannot see, as plainly as others do, & as I sh^{d} wish to do, evidence of design & beneficence on all sides of us. There seems to me too much misery in the world'.[66] Darwin's former belief in the ultimate justice and benevolence apparent in the laws of nature was additionally challenged by the death of his favourite daughter, Annie.[67] In his *unexpurgated Autobiography*, Darwin (although still wavering) conceded that the 'old argument from the existence of suffering against the existence of an intelligent first cause seems to me a strong one; whereas ... the presence of much suffering agrees well with that all organic beings have been developed through variation and natural selection'.[68]

After all of these struggles, even Darwin's *Autobiography*, which was also intended for Emma, reveals that, by 1876, he had lost at least any firm belief in an omniscient benevolent God and called himself an agnostic.[69] Although Darwin differed from most anti-religious thinkers in not taking any pleasure in reviling religion, Professor Ghiselin's remark that 'an agnostic is an atheist with children' appears to fit well here – 'and a pious wife' is all I add to it.[70] Darwin's pious transformation of natural theology into natural selection had finally undermined his previous firm belief in Christianity.

---

[62] Cf. J. Moore (1989), 222. A. Desmond and J. Moore (1991), xvi, cf. also 41, 281, 287, 354, 516, 519, 524, 531, 73, 677; cf. e.g. Peckham, ed. (1959), ch. VI, 241.

[63] See N. Barlow, *Charles Darwin's Ill-Health* (1958), in Barlow, ed. (1876), 240–43.

[64] Barlow, ed. (1876), 91, 44, 138.

[65] Peckham, ed. (1959) (variorum edition), ch. III, s. 165, ch. XIV, s. 269.

[66] Burkhard, et al., eds (1985), vol. 8, Letter to Asa Gray, 22nd May [1860], 224.

[67] J. Moore (1989), 216–20, 223, lays more stress on these personal reasons for Darwin's loss of faith. See lso J.H. Brooke, 'Darwin and Victorian Christianity', in J. Hodge and G. Radick, *Cambridge Companion to Darwin*, Cambridge, 2003, 199–202.

[68] Barlow, ed. (1876), 90, cf. 88–9.

[69] *Ibid.*, 94; Letter to J. Fordyce, 7th May 1879, in Burkhard, et al., eds (1958), 12041, or Darwin, ed. (1897), vol. 1, ch. VIII, 274. But cf. also B. Lightman, 'Huxley and Scientific Agnosticism', *British Journal for the History of Science* (2002), **35**, 271–89.

[70] Private discussion with Professor M. Ghiselin at a conference on *Naturphilosophie nach Schelling* at the University of Jena, Nov. 2002. Cf. also Barlow, ed. (1876), 93 and Darwin, ed. (1897), vol. 1, 286: F. Darwin mentions that Dr Aveling presents 'quite fairly' his father's views by stating that the terms 'agnostic' and 'atheist' were practically equivalent and agnostics only expressed disbelief in a much less aggressive way.

The Possibility of Biological Compromises Based on Lost Religious Grounds

As it no longer appeared possible to associate 'natural selection' with the invisible hand of a good, omniscient deity and the paradise-like Paleyan harmony of nature, Darwin's biological concept of the universality of natural selection and adaptation became weakened. Nevertheless, while Darwin's religious beliefs withered, he became increasingly free to modify his biological approach, which was still the object of strong scientific and public criticism.[71] Moreover, as Darwin uncompromisingly had aimed to apply his theory of natural selection to the *Descent of Man*, he, an exceptionally humane man, was apparently shaken by the concept of a 'remorseless struggle'.

Although still advocating individual natural selection as the main force of evolution, by the time he wrote the *Decent of Man* Darwin gave more room to sexual selection, group selection, correlation of growth and use or disuse. He explicitly conceded that he 'perhaps attributed too much to the action of natural selection or the survival of the fittest' and overestimated the omnipresence of adaptation.[72] Correspondingly, Darwin also altered the fifth and sixth edition of the *Origin*.[73]

With regard to his earlier pan-adaptationism, Darwin conceded in the *Descent of Man* that he 'did not formerly consider sufficiently the existence of structures, which, as far as we can at present judge, are neither beneficial nor injurious, and this I believe to be one of the greatest oversights as yet detected in my work'. He himself attributed his former bias in adopting a pan-adaptationist view of nature to his earlier belief in natural theology: 'I was not, however, able to annul the influence of my former belief, then almost universal, that each species had been purposely created; and this led to my tacit assumption that every detail of structure, excepting rudiments, was of some special, though unrecognised, service'. With regard to the universality of natural selection, Darwin conceded, after referring to Paley's influences on his thoughts, that anyone with the assumptions influenced by theology he had had in mind, would 'naturally extend too far the action of natural selection ...'.[74]

---

[71] R. Owen, G.D. Campbell (Duke of Argyll), G. Mivart, S. Butler and W. Thompson (Lord Kelvin) were untiring in pointing out problematic aspects of Darwin's theory. Also friends and supporters of Darwin, like A. Gray, C. Lyell, E. Haeckel and even T.H. Huxley and A.R. Wallace were far from advocating pure Darwinian dogma. P. Bowler, *The Non-Darwinian Revolution*, Baltimore, MD, 1988, esp. 5, 16, 76 f., 105–7, 175. Also Bowler (1990), e.g. 81, 155–61, 166.

[72] Barrett, et al., eds (1986–89), *Descent of Man* (1877), ch. II, orig., 61, cf. footnote 74.

[73] Peckham, ed. (1959), e.g. ch. I, 322:f; ch. III, 3:e; ch. IV, 95.14:e, f, 220:f; ch. XIV, 183:e, f, 183.0.0.1:f.

[74] The quotations in this paragraph are taken from Barrett, et al., eds (1986–89), *Descent of Man* (1877), ch. II, orig., 61; cf. also orig., v, 30 f., 57. Although Darwin allows for more causal pluralism, he still seems to emphasise that the 'so-called laws apply equally to man and the lower animals; and most of them even to plants', orig. 30.

On the Self-Undermining Dynamics of Ideas Between Belief and Science in the History of Ideas

If my abridged account of the development of Darwin's thought in relation to belief and science is correct, it may shed light on the more general understanding of the history of ideas. Taking the example given here, it can be seen that science and belief are neither uncoupled nor simply in opposition, but are interwoven, mutually influencing and undermining each other and, indirectly, even undermine themselves. From this point of view, Darwinism is neither a particular form of 'Christian' nor of 'anti-Christian biology', but – paradoxically – both. This essay shows how the inconsistency of accounts that stress either cooperation or conflict in Darwin's scientific and religious ideas can be resolved by understanding Darwin's intellectual development as a dynamic processes in which there was a self-undermining of ideas. Moreover, this case study of the eminent scientist Charles Darwin demonstrates the inadequacy or incompleteness of conventional understandings in the history of science. I would like to mention a few potential historiographic implications deriving from the account given.

Firstly, the conflict account is at odds with approaches exclusively based on empiricism, positivism and internalism. Without intending to deny the relevance of Darwin's empirical investigations to his theoretical achievements, it has been shown above that the construction of his theory also essentially hinges on religious or metaphysical tenets. This is consonant with the fact that Darwin did not adopt his theory of natural selection while staying on the Galapagos Islands, but when he was crystallising his impressions or his *protocol sentences* in the light of the available generalisations of that time.[75] I have shown that Darwin built his specific theory of evolution based on metaphysical tenets, which initially appeared to him to have a strong ethical and religious appeal. Moreover, I have shown that Darwin's resulting loss of faith, in turn, enabled him to modify his former biological theory. In pursuing a view of science which also considers metaphysical and religious aspects, I have advanced a historiographic position which has been influenced by the works of Professor Dr David Knight, in whose honour this anthology was written: 'Science is, and always was, based on a judicious mixture of empiricism and faith'.[76]

Secondly, my account differs from approaches that regard ideas and metaphysical aspects of theories as epiphenomenal superstructures based exclusively on the egoistic interests of the holder of the ideas. Nonetheless, my argument above suggests that

---

[75] At least since Hume, it has become apparent that generalisations – by definition – transcend the empirically given. I have treated the problem of induction elsewhere. Without following Popper in his denial of any positive solution of the problem of induction, I also regard generalisations as empirically under-determined and laden with theory.

[76] D.M. Knight, *The Age of Science: the Scientific world-view in the nineteenth century*, Oxford, New York, 1986, 70.

socio-economic conditions and interests may well play an important role in the history of science. The political background of Darwin, a Whig, may indeed have driven him to adopt his Malthusian explanation of evolution and to make nature an ally of the Victorian middle classes and their interest in a society based on free competition.[77] But, this is not the whole story. I have shown that the dynamics of self-undermining ideas were *also* important in the development of Darwin's thought. The end result was that he was forced to give up ideas which he would formerly liked to have maintained. This, I think, supports the view of the historian J.C. Greene, who once said, 'I still believe that, in some sense, human beings transcend nature. If ideas are only manifestations of a class interest or libidinal drives, then the whole intellectual enterprise is reduced to absurdity'.[78]

Thirdly, the above account differs from the approach of Kuhn, who holds the view that paradigms are largely incommensurable and disconnected. Because of this, Kuhn has been accused of assigning fundamental theoretical change to the irrational.[79] Although this case study follows Kuhn in stressing the importance of fundamental metaphysical changes, at a biographical level, here it has been shown that the succeeding 'paradigms' in Darwin's intellectual development were nevertheless closely, and in some sense rationally, connected.

Finally, the account given here stands in contrast with the approaches of K.R. Popper, D.T. Campbell, S. Toulmin and D.L. Hull which hold that the development of science can be attributed to self-preserving ideas in a process which is one of mere trial and error of conjecture and refutation.[80] Instead, it has been shown that a synthesis of ideas can gain tendencies lacking in those ideas themselves, that the dynamics of ideas follow an inner logic, and that ideas may even undermine themselves. Overall, this essay has emphasised the importance of the structural dynamics of ideas bridging belief and science and the ways in which ideas even rationally undermine themselves. This view, in some respects, appears to resemble the concept stemming from antiquity of an emanating, developing and rationally unfolding *logos*, which – perhaps – in other words may be expressed as the unfolding of a rational spirit.

---

[77] A. Desmond and J. Moore (1992/91), particularly 414; A. Desmond, *The Politics of Evolution*, London and Chicago, 1989.

[78] J.C. Greene, Introductory Conversation in J.R. Moore , ed. (1989), 10, cf. 11. Cf. J.C. Greene, *Science, Ideology, and World View*, Berkeley, etc., 1981, 1–2.

[79] Cf. T. Kuhn, *Die Struktur wissenschaftlicher Revolutionen*, orig., *the Structure of scientific revolutions*, Frankfurt a. M. 1991/62, e.g. 19, 116.

[80] K.R. Popper, *Objective Knowledge. An Evolutionary Approach*, Oxford, etc., 1972/79; D.L.Hull, *The Metaphysics of Evolution*, Chicago, 1988. For further developments of 'process-Darwinism' and its critique e.g. M. von Sydow (2001).

Chapter 11

Michael Faraday Meets the 'High-Priestess of God's Works': A Romance on the Theme of Science and Religion

Geoffrey Cantor

Introduction

In his contribution to *Faraday Rediscovered: Essays on the Life and Work of Michael Faraday, 1791–1867*, David Knight reflected on the relationship between Faraday and his mentor Humphry Davy. Although both men were from humble backgrounds, David noted their very different attitudes to and involvement in contemporary polite society. The major differences are implicit in his claim that Faraday 'married an *hausfrau* rather than an heiress, and refused Presidencies and titles'.[1] In contrast to the status-seeking Davy, with his penchant for heiresses, high society and honours, Faraday lived simply and was not driven by social ambition. While Davy married Mrs Jane Apreece, widow of Shuckburgh Ashby Apreece, first baronet of Washingley, Faraday, the son of an ironsmith, married Sarah Barnard, the daughter of a silversmith. Moving from iron to silver represented an elevation in affluence but not a great social advance, since both his father and his father-in-law were skilled artisans. An early pencil sketch possibly by Faraday himself shows Sarah to have been rather plain (Figure 11.1). A later photograph (dating from the 1840s) reinforces our impression of her as a *Hausfrau* who did not seek a place in high society (Figure 11.2). She doted on her husband, gave him emotional support but never sought the limelight. Yet the term *Hausfrau* does not capture her many sterling qualities. She was a pious Christian and her letters show her to have been a strong and independent person; while Faraday's letters in turn indicate his love and delicate feelings towards her.

There were many other women in Faraday's life. Except in certain male preserves, such as the Royal Society, he was surrounded by women, most of whom might also be

[1] David M. Knight, 'Davy and Faraday: Fathers and sons', in David Gooding and Frank A.J.L. James, eds, *Faraday Rediscovered: Essays on the life and work of Michael Faraday, 1791–1867*, Basingstoke, 1985, 33–49, 35.

Figure 11.1 **Pencil sketch of Sarah Barnard, c.1820, probably by Faraday. Reproduced by courtesy of the Royal Institution of Great Britain.**

Figure 11.2 Daguerrotype of Sarah and Michael Faraday dating from the mid-1840s. Reproduced by courtesy of the Royal Institution of Great Britain.

termed *Hausfrauen*, since they did not belong to the elevated ranks of society. He was frequently in the company of his mother, who lived to the age of 74, his two sisters, and the various unmarried nieces who cared for the ageing Faradays. Faraday belonged to a primitive Christian sect called the Sandemanians and at the meeting house women far outnumbered men in the ratio two to one.[2] Moreover, Faraday considered nature to be female, describing her as a 'goddess'. But this 'goddess', whose company he frequently sought, was rather asexual and not in the least motherly.[3]

While in his private life he was surrounded by worthy women, his public career in science brought him into contact with women from the higher echelons of society. For example, soon after joining the Royal Institution he travelled on the Continent with Davy and Davy's wife – the 'heiress' whom we have already encountered. Although Humphry Davy treated Faraday as a junior member of the international community of scientists, Jane Davy openly despised him and treated him as a lowly valet; while Faraday, in turn, felt deeply wounded by this snobbish ornament of fashionable society. Subsequently, his audience at the Royal Institution included many society women who were definitely not *Hausfrauen*, possibly like those portrayed in an 1857 *Punch* cartoon (Figure 11.3).[4] As George Eliot remarked, for society ladies Faraday's lectures were 'as fashionable an amusement as the Opera'.[5] Faraday's relationship with one such fashionable lady is the subject of this paper.

This lady was the recipient of a letter in which Faraday made the frequently cited assertion: 'There is no philosophy in my religion'.[6] Faraday was always precise in his use of language and this short but concise statement therefore deserves our close attention. He seems to be saying that he did not derive any scientific information or assumptions from his religion ('philosophy' being roughly equivalent to our term 'science'). Instead, he maintained a strict separation between his science and his religion. In the terminology of Ian Barbour, who has written on the relationship between science and religion, Faraday asserted their *Independence*.[7] This interpretation gains some credence if we examine a later passage from the same letter:

---

[2] Geoffrey Cantor, *Michael Faraday, Sandemanian and Scientist*, Basingstoke, 1991, 268–71.
[3] Henry Bence Jones, *The Life and Letters of Faraday*, 2nd edn, 2 vols, London, 1870, i, 267.
[4] 'Mary Ann's Notions', *Punch* (1857), 32, 109.
[5] G.S. Haight, ed., *The George Eliot Letters*, London, 1954, i, 241–2.
[6] Faraday to Augusta Ada Lovelace, 24 October 1844, in Frank A.J.L. James, ed., *The Correspondence of Michael Faraday*, London, 1996, iii, 264–6.
[7] Ian G. Barbour, *Religion and Science: Historical and contemporary issues*, San Francisco, 1997, 84–9. See also Geoffrey Cantor and Chris Kenny, 'Barbour's Four-Fold Way: Problems with his taxonomy of science–religion relationships', *Zygon* (2001), 36, 763–79.
[8] Faraday to Lovelace, 24 October 1844, in James, ed. (1996).

MARY ANN'S NOTIONS.

MY DEAR—no, I shan't,

"I NEVER knew such a cross old unkind good-for-nothing old thing as you are in all my life. I was beginning to be quite friendly with you, and to write to you with confidence, and then you suddenly turn snappish and sulky, and put such a note as that to my letter as you did last time. I know very well what it was about. I made a little mistake, and mixed up the LORD CHANCELLOR LEWIS with LORD CRANWORTH, the Chancellor of the Exchequer. Why could not you have set me right, and what is the use of printers and all those sort of people [1] if they cannot correct little inadvertencies like that? And then for you to put a cross note, and threaten to end our correspondence, I thought you were so old as

Figure 11.3 **Like these fashionable admirers, Mary Ann was attracted to 'dear' Michael Faraday and described his lecture as 'lovely'. *Punch* (1857), 32, 109. Reproduced with the kind permission of Leeds University Library.**

> I do not think it at all necessary to tie the study of the natural sciences & religion together and in my intercourse with my fellow creatures that which is religious & that which is philosophical have ever been two distinct things.[8]

This, surely, is a definitive statement: Faraday kept his science and religion firmly apart. However, there appear to be two rather different claims in this last quotation. In the first half Faraday denies the need to conjoin science and religion. This is a rather weak claim since all it asserts is that both science and religion can stand independently; one does not require the other for support. Thus a person could be religious and ignore the whole of modern science or vice versa. From the impersonal phrasing of this quotation we should not conclude that Faraday himself separated science and religion. Indeed, the first part of the sentence allows the *possibility* that someone – even Faraday – might wish to tie science and religion together.

The latter part of this quotation can now be used to clarify Faraday's intention; in his intercourse *with other people* he drew a clear separation between his religion and his science. This position seems to cohere with Faraday's well-attested reluctance to talk to others about his religious beliefs. For example, his friend and biographer John Tyndall noted: 'Never once during an intimacy of fifteen years did he mention religion to me, save when I drew him on to the subject.'[9] This statement of Tyndall's confirms the non-missionary and almost secretive way Sandemanians practised their religion and their rejection of other forms of Christianity.

We now seem to possess a clear interpretation of the text: in his public pronouncements Faraday maintained a strict separation between science and religion. This emphasis on context should be noted and it is certainly true that Faraday did not allude to religion in the papers he presented to the Royal Society. By contrast, religious discussions sometimes occurred in his private letters (and then only when writing to fellow Sandemanians and to a few sympathetic correspondents).

However, the above interpretation encounters a major problem if we look instead at the lectures he delivered at the Royal Institution, where his audience usually consisted of several hundred men and women holding a variety of religious commitments. Far from eschewing religion, Faraday often introduced religious issues as being integral to the science content of these lectures.

One of the most striking examples occurs in a course of lectures on 'Electricity and Magnetism', which he delivered in the mid-1840s – a period when he was particularly exercised by the religious implications of his natural philosophy. Throughout this series he frequently raised religious issues which show that his ideas about forces (which he often called 'powers') and the structure of matter were deeply infused with religious meaning. For example, he opened this 1846 series by alerting his audience to 'the power[s] with which the Creator has gifted … matter'. One such power was

[9] John Tyndall, *Faraday as a Discoverer*, New York, 1961, 180.

gravitational attraction, but Faraday was particularly interested in the electric, magnetic and chemical powers – these were the principal subjects of his scientific researches. It is also important to note that these powers were subject to laws – another scientific term that he (and others) endowed with religious meaning: 'Such is the beauty of the manifestations that come before us when we deal with the great laws of nature, and observe phenomena which never fail, because the Creator governs by fixed laws.'[10] Later in the series he commented on the 'great system of nature, which changes not; that all its laws were established from the beginning ... [they] are as *old* as creation'.[11]

Many other examples can be cited of public occasions when Faraday clearly mixed science and religion. These show that, contrary to my earlier interpretation of the text, he was patently *not* maintaining the *Independence* between science and religion. So far in this paper I have signally failed to illuminate Faraday's letter and especially the crucial sentence: 'There is no philosophy in my religion'. The line of argument I have pursued is manifestly faulty. Why have I gone astray?

My mistake is that I forsook the advice of my teachers – and the advice I give to my students – to pay attention to *context*. Instead I have concentrated solely on the text. To begin to re-evaluate the text I shall raise the issue of context by framing a few questions. Who was Faraday addressing in the letter from which I have cited? When was it written? What events led Faraday to write this letter? What was he trying to achieve in this letter? The answer to the first of these questions is straightforward: Faraday's correspondent was Ada, the Countess of Lovelace.

Faraday's Correspondent: Ada, Countess of Lovace[12]

She was the daughter of Lord Byron. Imaginative, beautiful and highly intelligent but bored by a conventional marriage and the constraints of polite society, she continually sought exciting outlets for her scientific talents (Figure 11.4). She took a keen interest in science and especially in Charles Babbage's attempts to construct his analytical engine. In 1843 she published her translation of Luigi Federico Menabrea's memoir on Babbage's machine, adding extensive notes of her own.[13] To aid her explanation she

---

[10] Michael Faraday, 'A Course of Lectures on Electricity and Magnetism', *London Medical Gazette* (1846), 3, 1–7, 4. On Faraday's theory of powers see Cantor (1991), 174–8.

[11] Faraday (1846), 523–9, 523.

[12] On Ada Lovelace see Doris L. Moore, *Ada, Countess of Lovelace: Byron's legitimate daughter*, London, 1977; Dorothy Stein, *Ada. A Life and a Legacy*, Cambridge, MA, 1985; Betty A. Toole, *Ada, the Enchantress of Numbers: a Selection from the letters of Byron's daughter and her description of the first computer*, Mill Valley, CA, 1992.

[13] L.F. Menabrea, 'Sketch of the Analytical Engine Invented by Charles Babbage ... with notes upon the memoir by the translator A.A.L.', *Taylor's Scientific Memoirs* (1843), 3, 666–731.

Figure 11.4 **Ada Lovelace (1815-52), based on a portrait by Alfred Edward Chalon, R.A., and published in Bentley's Miscellany (1853), 33, 69. Reproduced with the kind permission of the Leeds Library.**

drew the comparison between the analytical engine and a Jacquard Loom, which can be programmed to weave a pattern. She thereby participated in one of the most exciting scientific projects of the day. Her infatuation with Babbage ended in the summer of 1844 and she cast her gaze in the direction of Faraday. This leads us to the second question – when was the letter written? It is dated 24 October 1844. Interestingly, Faraday received it when he had largely recovered from his 'breakdown' – a period when he was particularly reflecting on the religious and philosophical implications of science.[14]

Lovelace probably first encountered Faraday when she attended his lectures at the Royal Institution. Although she may have talked to him at the Royal Institution or elsewhere, the first solid evidence of communication dates from a few months earlier when he sent her a copy of his paper entitled 'A Speculation Touching Electric Conduction and the Nature of Matter', which appeared in the *Philosophical Magazine* for February 1844. In this fascinating and significant paper Faraday argued against the received view that matter consists of small, hard, billiard-ball-like particles moving in a vacuum. Instead, drawing on the ideas of Roger Boscovich and Joseph Priestley he suggested that matter can be reduced conceptually to fields of force. In this paper Faraday confronted a major and profound question, and marshalled a variety of empirical and metaphysical arguments.[15] This shows Faraday at his most creative and his mental gymnastics must have captivated the intellectually precocious Lovelace.

In her spirited reply of 16 October 1844 Lovelace forsook traditional Victorian reserve and threw herself at Faraday, begging him to allow her to become his acolyte. Before turning to the content of this letter we should first note that she had already developed close but often strained friendships with other members of the scientific community. In 1843 the physiologist William Carpenter, who tutored Lovelace's children, tried to become her intimate confidant and healer, but, as Alison Winter has noted, conventional social mores, accentuated by the difference in social status, and the threat to Carpenter's reputation resulted in a more formal and distant relationship.[16] In a later letter to her mother she expressed her tendency to exert 'the powers I *know I have over others*' and also her conviction that she was destined to play a divinely ordained role in revealing great truths about God's creation. While attempting to moderate her personal influence over others she wished to 'continue to be simply the High-Priestess of Babbage's Engine, & serve my apprenticeship faithfully therein, before I fancy myself worthy to approach being the *High Priestess* of God

---

[14] On Faraday's 'breakdown' see James F. O'Brien, 'Faraday's Health Problems', *Bulletin for the History of Chemistry* (1991), 11, 47–50.

[15] Michael Faraday, 'A Speculation Touching Electric Conduction and the Nature of Matter', *Philosophical Magazine* (1844), 24, 136–44.

[16] Alison Winter, 'A Calculus of Suffering: Ada Lovelace and the bodily constraints of women's knowledge in early Victorian England', in Christopher Lawrence and Steven Shapin, eds, *Science Incarnate: Historical Embodiments of Natural Knowledge*, Chicago, 1998, 202–39.

Almighty Himself'.[17] Also, on 11 November 1844, less than three weeks after the letter from Faraday, she informed her mother that Faraday considered her 'the *rising star* of Science … I may be the Deborah, the Elijah of *Science*'.[18] Her grounds for reaching this conclusion are unclear; certainly the measured tone of Faraday's letter of 24 October did not justify it.

Extensive quotations are given from the letter Augusta Ada Lovelace wrote to Michael Faraday on 16 October 1844:

> You will be kind enough to think of me simply as one of *God's children*. The mere accidents of my being an inhabitant of this particular planet, of this particular corner of it[,] England, & of my wearing the *female form*, (with a human coronet to boot stuck at the apex), *these* constitute only one particular case of the *general formula* in which God has chosen to create moral beings fitted to hold relations with each other, & with *Him*.
>
> I have long been *vowed to the Temple*; – the Temple of *Truth, Nature, Science*! And every year I take vows more strict, till now I am just entering those portals & those mysteries which cut of[f] all retreat, & bind my very life & soul to *unwearied & undivided* science at it's altars henceforward. I hope to die the *High-Priestess* of God's works as manifested on this earth, & to earn a right to bequeath to my posterity the following motto, '*Dei Naturaeque Interpres*' [Interpreter of God and nature].
>
> Lovelace then begged Faraday to allow her to 'become in *some* respects your disciple'.
>
> … I hope to crystallise before my death into a radiant & burning star for the light of mankind.
>
> One reason why I desire to become more intimate with you, is my opinion of your *moral & religious* feelings. –
>
> Do *you* agree with me, as I rather expect you do, in the impression that the *highest & most penetrating* degree of *intellect, that* species of it which is alone fitted to deal with the more *subtle and occult* agents of nature, is unattainable excepting thro' a high *spiritual and moral* development; far higher than it is usual even to *aim* at. *I* believe there is a connexion between the two that is not understood or suspected by mankind.
>
> *You* are the only philosopher [≈ scientist] I have ever seen, who gave me the impression of feeling this in it's full force …
>
> I do not know to what particular *sect* of Christians you belong, or whether to *any*; nor do I think that much matters.
>
> I am myself a Unitarian Christian; as far as regards some of their views of Christ that is.
>
> But in truth, I cannot be said to be anything but *myself*. In *some* points I am a *Swedenborgian* in feelings. Again in others I am slightly Roman Catholic; & I have also my alliance with the older *Rosecrucians*.[19]

---

[17] Lovelace to Lady Byron, 15 September [1843], in Toole (1992), 264–5. On Lovelace's involvement with mesmerism see Alison Winter, *Mesmerized. Powers of Mind in Victorian Britain*, Chicago, 1998, 230–33.

[18] Lovelace to Lady Byron, 11 November 1844, in Toole (1992), 290–92.

[19] Lovelace to Faraday, 16 October 1844, in James, ed. (1996), 253–5.

Given the lack of opportunity available to women to pursue science during the mid-nineteenth century, it is interesting that Lovelace specified her role in distinctly religious terms as a prophet and 'High Priestess'. She was clearly not prepared to sweep the laboratory or learn the skills of glass blowing or volumetric analysis. Nothing so mundane! – that might be suitable for a *Hausfrau* but not herself. She knew what she wanted and believed that Faraday could assist her in achieving her goal.

How did Faraday respond? According to James Hamilton: 'Before Ada Lovelace's epistolary glare, Faraday was as a rabbit in front of a snake'.[20] Although this assessment may partially capture the tone of their correspondence, we must be much more precise. Faraday would have been shocked and embarrassed by her over-familiarity, which he had probably not encouraged. Unless there were other relevant events about which we possess no evidence, this letter seems a remarkable overreaction to his 'A Speculation Touching Electric Conduction and the Nature of Matter', sent a few weeks earlier. But he would also have been aware of the vast age difference – 24 years – and the social gulf that separated them. She was married to William King, eighth Baron King and Baron Ockham, Lord Lieutenant of Surrey, who was in 1838 created Earl of Lovelace. She moved effortlessly in high society, counted among her acquaintances members of the aristocracy as well as scientists like Babbage, Faraday, Carpenter, Augustus de Morgan, David Brewster, and Charles Wheatstone. By contrast, he was an ironsmith's son and a member of a dour religious sect.

Faraday frequently came across members of the aristocracy at the Royal Institution but it is doubtful whether he had previously encountered a female aristocrat who made such personal demands. She was the very antithesis of the pious *Hausfrau* to whom he was married and the other female members both of his family and of the Sandemanian church. He was both frightened of her and doubtless flattered by her charm and attention. Clearly aware of her superior social standing and of her femininity, he replied with characteristic politeness and gentility. In his letter of 24 October after some vague but encouraging opening comments he pleaded ill-health in order to justify his ten-day delay: 'That I should rejoice to aid you in your purpose you cannot doubt ... [but] I am a labourer of many years' standing made daily to feel' worn out. His body was becoming increasingly weak, he complained. His memory frequently failed him and he often felt giddy. His doctor had repeatedly advised him to 'retire to the seaside and to inaction'. While there was some truth in these excuses, Faraday also deployed them to help maintain his distance from her and avoid her over-zealous demands. Unsurprisingly he declined to accept her as a student – he simply could not have coped.

Finally, Faraday had to confront the religious views that Lovelace had expounded: 'I am myself a Unitarian Christian; as far as regards some of their views of Christ that is. ... In some points I am a *Swedenborgian* in feelings. Again in others I am slightly

[20] James Hamilton, *Faraday. The Life*, London, 2002, 311.

Roman Catholic; & I have also my alliance with the older *Rosecrucians*.' Significantly he avoided commenting on her statements, instead articulating his own position, but in a very abbreviated form that succeeded in concealing his deeper religious commitments. Indeed, he prefaced his comments by stating that Lovelace 'will be sadly disappointed' in his religious views.[21] He is, he says, a devout primitive Christian whose hope is founded 'on the faith that is in Christ'. Implicitly he repudiated Lovelace's Unitarianism – which stood opposed to his evangelical faith – and her Catholicism.[22] Likewise he would have dismissed as unacceptable all forms of nature worship such as the mystical teachings of both the Swedenborgians and the Rosicrucians, who held that nature is thoroughly spiritualised and that we are transcendent beings able to become god-like. From Faraday's perspective this incomprehensible farrago of science and religion was thoroughly unacceptable. Nature was God's creation and not itself divine. One can hardly imagine a mix of religious ideas less palatable to Faraday.

He would also have been repelled by another theme expressed by Lovelace. Sandemanians reject any form of priesthood, not only the pastors of the Catholic Church, universally abhorred by Protestants, but also the clergy of the Established Churches. For Sandemanians, who broke away from the Church of Scotland early in the eighteenth century, one of the main points of contention concerned church structure, the Sandemanians insisting that there was no biblical warrant for either a professional ministry or the Presbyterian bureaucratic structure. Instead they insisted that true Christians must read the Bible, without the interference of any mediator, and live according to its teachings.[23] Faraday would therefore have dismissed as

---

[21] The text of the final paragraph of Faraday's letter of 24 October 1844 [James, ed. (1996), 264–6] reads:

> You speak of religion & here you will be sadly disappointed in me. You will perhaps remember that I guessed[,] & not very far aside[,] your tendency in this respect. Your confidence in me claims in return mine to you[,] which indeed I have no hesitation to give on fitting occasions[,] but these I think are very few[,] for in my mind religious conversation is generally in vain. There is no philosophy in my religion[.] I am of a very small & despised sect of christians known, if known at all, as *Sandemanians*, and our hope is founded on the faith that is in Christ. But though the natural works of God can never by any possibility come into contradiction with the higher things that belong to our future existence, and must with every thing concerning Him ever glorify him[,] still I do not think it at all necessary to tie the study of the natural sciences & religion together and in my intercourse with my fellow creatures that which is religious & that which is philosophical have ever been two distinct things.

[22] For Faraday's antipathy towards Catholicism see his comments from Rome in 1814, in Bence Jones (1870), i, 153–60.

[23] On the history and doctrines of the Sandemanian sect see Cantor (1991), 11–36; Derek B. Murray, 'The Influence of John Glas', *Records of the Scottish Church History Society* (1984), 22, 45–56; Lynn McMillon, *Restoration Roots*, Dallas, 1983.

unacceptable Lovelace's self-appointed role as 'High-Priestess'. In pursuing science he had direct access to the Book of Nature and his reading of it was not dependent on the mediation of any priest or priestess.

We are now in a position to appreciate other aspects of his response on 24 October. Always the gentleman, he answered politely but allowed her no possible riposte. He felt obliged to reply to her effusive comments on religion, but was pointedly brief, claiming that discussion of religious matters is generally fruitless. Immediately after this he asserted: 'There is no philosophy in my religion'. He wrote this, I suggest, in order to disarm Lovelace's pantheistic strategy of integrating science with her wild religious notions. Rather than run the risk of offending her by stating outright his opposition to her views he sought to imply that science and religion should not be mingled in the way she had indicated in her letter. Moreover, in order to distance himself even further from her views he went on to assert: 'I do not think it at all necessary to tie the study of the natural sciences & religion together and in my intercourse with my fellow creatures that which is religious & that which is philosophical have ever been two distinct things'.

He was employing the strategy of *Independence* against Lovelace. Given her lack of propriety in religious matters he was implying that he refused to accept her heady mix of science and religion. Science and religion should not be tied together in the way she attempted. This does not imply that science and religion are completely separate; rather, one is not dependent on the other.

The meaning of the latter part of the preceding quotation also becomes clearer when we remember his reticence on discussing Sandemanianism in the presence of (non-Sandemanian) scientists – see the Tyndall quotation above. Likewise he would not have talked about lines of force at the Sandemanian meeting house, where members of the sect congregated to read and meditate on God's inspired words. Each activity was appropriate to its own social environment.

We must now address one final passage in Faraday's letter of 24 October in which he stated that 'the natural works of God can never by any possibility come into contradiction with the higher things that belong to our future existence, and must with every thing concerning Him ever glorify him'. This can likewise be read as a riposte to Lovelace's pagan and pantheistic beliefs. In contrast to her views, he drew a sharp distinction between God's works and the revealed truths contained in the Bible concerning our afterlife. His arguments are thoroughly conventional. First, God's works cannot contradict the spiritual message taught by Christ, and, second, they display God's providence.

Following Faraday's letter of 24 October, a few letters passed between Faraday and the countess. As some of these letters have not survived we can only partially reconstruct their later interactions. Suffice to say that, despite his refusal, she continued to try to persuade him to become her mentor in science. It is not clear whether his letter of 24 October had been too encouraging or whether she was simply encouraged by his courteous reply. Finally, however, he told her unequivocally that he was being driven mad by her badgering and could not meet her again.[24] The correspondence ended soon thereafter.

Before returning to Faraday we should empathise with Lovelace's situation. Here was a highly intelligent and highly spirited woman trapped in a conventional and intellectually unsatisfying marriage. Moreover, despite her social situation – perhaps because of her social situation – her life was circumscribed and the possibilities of participating in science were limited. She desperately sought an outlet for the immense well-spring of her creative energy by engaging with exciting contemporary developments in science. She mixed with many members of the scientific community and was particularly attracted to some of its more progressive spirits, like de Morgan, Babbage and Faraday.

Faraday was far less help to her than were de Morgan or Babbage. He could certainly see her predicament but failed to respond as she wished. The gap was too great, both socially and religiously. Inhibited by Victorian social conventions and his conservative religious views he tried to sound moderately encouraging but remained defensive. I therefore end by suggesting that the oft-quoted sentence, 'There is no philosophy in my religion', should be read in the context of Lovelace's earlier letter and not as a well-formulated claim about Faraday's own position on the relationship between science and religion. Moreover, this apparently specific statement is of little assistance to us in determining Faraday's own views about the relation between science and religion, an issue that he considered very important and on which he thought deeply.[25]

[24] Lovelace to Faraday, 24 October 1844, in James, ed. (1996), 267; Faraday to Lovelace, 26 October 1844, in *ibid.*, 270; Lovelace to Faraday, 27 October 1844, in *ibid.*, 270–72; Lovelace to Faraday, 8 November 1844, in *ibid.*, 276–7; Lovelace to Faraday, 10 November 1844, in *ibid.*, 279–81; Lovelace to Faraday, 13 November 1844, in *ibid.*, 282–3; Faraday to Lovelace, n.d. [late November 1844], in Hamilton (2002), 319–20; Lovelace to Faraday, 1 December 1844, in James, ed. (1996), 291–2.

[25] For Faraday's views on issues of science and religion see Cantor (1991).

Chapter 12

An 'Open Clash between Science and the Church'?: Wilberforce, Huxley and Hooker on Darwin at the British Association, Oxford, 1860

Frank A.J.L. James[1]

At about midday on Saturday 30 June 1860, in the library of the new University Museum in Oxford, between 400 and 1,000 men and women assembled to attend a hastily arranged meeting of Section D of the British Association. Those attending this session of the Zoology and Botany, including Physiology, Section became witnesses to an event that would become one of the major scientific icons of twentieth-century culture and one which still serves to epitomise the separation of scientific knowledge from religious belief which is such a characteristic feature of significant parts of contemporary British science and society.

That afternoon the Bishop of Oxford, Samuel Wilberforce (1805–73), then aged 54, one of the professors at the Royal School of Mines in London, Thomas Henry Huxley (1825–95), then aged 35, and the 43-year-old assistant director of the Royal Botanic Gardens, Kew, Joseph Dalton Hooker (1817–1911), with others, engaged in a discussion over *The Origin of Species* by Charles Darwin (1809–82) that had been published the previous November. The discussion was chaired by John Stevens Henslow (1796–1861), professor of Botany at Cambridge University and Rector of Hitcham.[2] He had been Darwin's mentor in Cambridge and, while not committed to

---

[1] I wish to thank the following for helpful information and comments on earlier versions of this essay: David Knight, Matthew Eddy, David Gooding, Helen Haste, William Gosling, Ruth Barton, Michael Roberts, Richard Bellon, Peter Bowler, Leonard Wilson and Michael Collie. I am also grateful for permission to study holdings in the following archives: Imperial College (IC), the Bodleian Library, the British Library, the Library of Congress, St Andrews University, Cambridge University Library (CUL) and the Académie des Sciences.

[2] For recent studies of these men see: Standish Meacham, *Lord Bishop: the Life of Samuel Wilberforce, 1805–1873*, Cambridge, Mass., 1970; Adrian Desmond, *Huxley: the Devil's disciple*, London, 1994; *idem.*, *Huxley: Evolution's High Priest*, London, 1997; Ray Desmond, *Sir Joseph Dalton Hooker: Traveller and Plant Collector*, Woodbridge, 1999; Adrian Desmond

the theory of natural selection, was not opposed to it and had indeed been seeking to neutralise hostility to the theory at the university.[3] As president of Section D in 1860, it fell to him to chair the session which would discuss Darwin's work.

This 1860 discussion is cited endlessly in popular and not so popular media. In terms of serious studies, scholars ranging from Stephen Jay Gould (1941–2002)[4] and J. Vernon Jensen[5] to John Brooke[6] and Janet Browne[7] and many others besides[8] have accorded it attention. In popular terms it was parodied at length in *The Water Babies* (1863) by Charles Kingsley (1819–75), briefly alluded to in that seminal inter-war historical text *1066 and all that*,[9] re-enacted in the 1978 BBC television series *The Voyage of Charles Darwin*,[10] was turned on its head in a 'replay' by the present Bishop of Oxford, Richard Harries, and the geologist Beverly Halsted (1933–91), when the British Association last met in Oxford in 1988[11] and a set of lectures on

---

and James Moore, *Darwin*, London, 1991; Janet Browne, *Charles Darwin: Voyaging*, New York, 1995; *idem.*, *Charles Darwin: the Power of place*, New York, 2002; S.M. Walters and E.A. Stow, *Darwin's Mentor: John Stevens Henslow 1796–1861*, Cambridge, 2001.

3 Henslow to Hooker, 10 May 1860, in Frederick Burkhardt, Sydney Smith, Duncan Porter, et al., eds, *The Correspondence of Charles Darwin*, 13 vols and continuing, Cambridge, 1985–, **8**: 200–201. See also 'Letter from Professor Henslow', in *Macmillan's Mag.* (1861), **3**, 336 and Browne (2002), 117–18.

4 Stephen Jay Gould, 'Knight Takes Bishop?', *Nat. Hist.* (1986), **95**(5), 18–33.

5 J. Vernon Jensen, 'Return to the Wilberforce–Huxley Debate', *Brit. J. Hist. Sci.* (1988), **21**, 161–79; *idem.*, *Thomas Henry Huxley: Communicating for science*, Newark DE, 1991, 63–86.

6 John Hedley Brooke, 'The Wilberforce–Huxley Debate: Why did it happen?', *Sci. Christian Belief* (2001), **13**, 127–41 is excellent on the background contexts.

7 Janet Browne, 'The Charles Darwin–Joseph Hooker correspondence: an Analysis of manuscript resources and their use in biography', *J. Soc. Bibliog. Nat. Hist.* (1978), **8**, 351–66, 361–2. Browne (2002), 113–25.

8 A.M. Armstrong, 'Samuel Wilberforce *versus* T.H. Huxley: a Retrospect', *Quart. Rev.* (1958), **296**, 425–37. J.R. Lucas, 'Wilbeforce and Huxley: a Legendary encounter', *Hist. J.* (1979), **22**, 313–30. Josef L. Altholz, 'The Huxley–Wilberforce Debate Revisited', *J. Med. Hist.* (1980), **35**, 313–16. Sheridan Gilley, 'The Huxley–Wilberforce Debate: a Reconsideration', in Keith Robbins ed., *Religion and Humanism: Papers read at the eighteenth summer meeting and the nineteenth winter meeting of the Ecclesiastical History Society*, Oxford, 1981, 325–40. Sheridan Gilley and Ann Loades, 'Thomas Henry Huxley: the War between science and religion', *J. Relig.* (1981), **61**, 285–308. Desmond (1994), 276–81.

9 Walter Caruthers Sellar and Robert Julian Yeatman, *1066 and All That. A memorable history of England. Comprising all the parts you can remember including 103 good things, 5 bad kings and two genuine dates*, London, 1930, 111 referring to '*Tails of a Grandfather*'.

10 The scene is described in Jensen (1988), 176. This re-enactment was certainly the spur for at least some of the articles that appeared in the years immediately following.

11 See Pearce Wright, 'Church Confronts Science: Evolution battle rolls on', *Times*, 10 September 1988, 4.

science and religion was delivered in the museum to mark its 140th anniversary in 2000.[12] Furthermore, an entire room at Darwin's recently refurbished home, Down House, is given over to the event, together with a reconstructed audio extract of the discussion; it is not clear, however, whether the non-expert visitor would appreciate that it was a reconstruction.

One of the reasons stories such as this become entrenched in the consciousness of society is because they do form a way of discoursing about contemporary concerns using apparently concrete historical examples. It is far easier to say that science and religion have been and are at 'war' by citing cases such as Galileo Galilei (1564–1642) or the Huxley–Wilberforce discussion, than by trying to disentangle the often complicated relations as they exist today. If such a straightforward polarised world existed, how would one discuss science and religion using John Habgood, Archbishop of York from 1983 until 1995, as an example embodying both? He, after all, at the Loughborough meeting of the British Association in 1994 suggested that monkeys might have souls, thus prompting the splendid headline in the *Daily Telegraph*: 'Apes have souls too, says primate'.[13] A view, which, as I will show, Wilberforce would have found repugnant.

The reason why the discussion that June afternoon continues to attract strong popular and scholarly interest is due to the discussion taking on a universal mythic significance. By the early twentieth century, along with the 1874 British Association Belfast presidential address by John Tyndall (1820–93),[14] the events of 1860 had come to be seen as one of the milestones in the process of the transformation of natural philosophy into natural science free from theological fetters. Leonard Huxley (1860–1933), writing right at the end of the nineteenth century, was quite explicit about this. The discussion, in his view, was an 'open clash between Science and the Church'[15] and that the 'importance of the Oxford meeting lay in the open resistance that was made to authority'.[16] So pervasive did this attitude become during the twentieth century that some scientists began to think that science had single-handedly destroyed the influence of Christianity.[17] This notion of the rise of science at the expense of Christianity also served the interests of churchmen and theologians since they could blame something else other than their own shortcomings which seem to me to be a more important factor in the decline of Christianity of Britain during the twentieth

---

[12] Which commenced with Brooke (2001).

[13] *Daily Telegraph*, 9 September 1994, 1

[14] Ruth Barton, 'John Tyndall, Pantheist: a Rereading of the Belfast address', *Osiris* (1987), **3**, 111–34.

[15] Leonard Huxley, *Life and Letters of Thomas Henry Huxley*, 2 vols, London, 1900, **1**, 181.

[16] *Ibid.*, **1**, 189.

[17] For a short discussion of science and religion in the twentieth century see the postscript in John Hedley Brooke, *Science and Religion: Some historical perspectives*, Cambridge, 1991.

century. Thus, supporting the myth simultaneously well served what might, at first sight, appear to be a number of divergent interests.

So how did the myth arise? In the remainder of this essay, I will provide an account of the discussion which shows how the individual protagonists, the sensibilities of the Gentlemen of Science of the British Association and the politics of Oxford, both church and university, all interacted to make it necessary for knowledge of the discussion to be, if not suppressed, at least sanitised. This laid the ground 20 years later for the myth to be created.

The history of the discussion was constructed in the final two decades of the nineteenth century by men (and one woman) to suggest that the conflict between science and religion then going on had commenced in the 1860s. The nature of this construction is perhaps best grasped by reflecting that so far no published description or indeed reference to the discussion (apart from the parodies) has been found from the end of 1860 until 1880 and there are comparatively few unpublished references. Thus the discussion that Jim Moore has said, after Waterloo, was the second best-known battle of the nineteenth century[18] has little history for the first 20 years of its life; although it has plenty after that. The same cannot be said of Waterloo fought in 1815. (Table 12.1).

Of those accounts written shortly afterwards, they fall into three categories most of which provide short descriptions of the discussion. These are newspaper and periodical reports, letters and diaries, and hearsay accounts. I have so far located ten contemporary letters and two diary entries written either by participants or by members of the audience which describe the discussion (Table 12.1), 14 newspaper or periodical reports (Table 12.2; note the accounts in the three Oxford papers are identical) and four hearsay reports (Table 12.3).

The most important of the contemporary accounts are the *Athenaeum* of 14 July 1860 (*B13*), the letters from the participants John William Draper (1811–82) (*A8*), Hooker (*A3*) and Huxley (*A12*) (although they each emphasise their respective contribution at the expense of the others) and the letter from Philip Pearsall Carpenter (1819–77) (*A10*) who was in the audience. The authorship of the printed articles, with the exception of Henry Fawcett (1833–84) in *Macmillan's Magazine* (*B14*), is anonymous. However, it is clear from Draper's letter (*A8*) that he was asked by the secretary of Section D, Edwin Lankester (1814–74) to write a few pages on his contribution for the *Athenaeum*. This raises the possibility that Lankester was responsible for arranging what is the longest contemporary account of the afternoon and would also explain why it was published a week after most of the other accounts.

---

[18] James Moore, *The Post-Darwininan Controversies: a Study of the Protestant struggle to come to terms with Darwin in Great Britain and America 1870–1900*, Cambridge, 1979, 60. See also Desmond and Moore (1991), 492 where the discussion is again compared with Waterloo.

Hardly anything was written about the discussion from 1860 until reminiscences and biographies of some of those involved in one way or another began to be published in the 1880s. During the next four decades accounts were published in 18 texts (Table 12.4). The length and detail varied widely, some just published a contemporary letter, some provided retrospective accounts, and for others letters of recollection were written especially for the purpose, some to be published (at least in part) and some to provide background information (Table 12.5).[19] The 20-year gap in published descriptions of the discussion is striking, and with only three intervening unpublished references (listed in Table 12.6) so far located, it is one that needs to be explained. Furthermore, with the passage of between 20 and 35 years since the discussion had occurred, these later accounts cannot, because of the fallibility of human memory, be relied on to provide irrefutable evidence of what happened. Each of these later accounts has some axe to grind as do the contemporary descriptions. Nevertheless the later accounts cannot be dismissed out of hand as sources and, at the very least, provide some evidence as to who was at the discussion.

So who was there? So far those in Table 12.7 can be identified from both contemporary and later sources as being present. As might be expected there was an overwhelming preponderance of Oxford men present. Also we can confirm the presence of a large number of Section D committee members. In terms of total participation the latter would have formed a far smaller proportion of those present at the discussion than implied here and so this list must not be taken as being statistically representative of those who were present.

All historical accounts of the discussion seem to take it for granted that this discussion arose spontaneously out of the deeply held convictions of those involved about the theory of natural selection. Wilberforce wrote in his diary that day 'In again to Sections & at Zoological called up by Henslow on Darwinian Theory & spoke at some length in controversy with Huxley' (*A1*) and again three days after the discussion he wrote to Charles Anderson (1804–91) that 'On Saturday Professor Henslow who presided over the Zoological Section called on me by name to address the section on Darwin's theory' (*A4*). It was almost as if he had not been expecting to speak. But the invitation for him to participate must have been agreed in advance by the Section Committee, since certainly Hooker was expecting him to speak (*A3*), as was John Richard Green (1837–83): 'On Saturday morning I met Jenkins going to the Museum. We joined company, and he proposed going to Section D, the Zoology, etc., "to hear the Bishop of Oxford smash Darwin"' (*A5*).

The *Literary Gazette* published on 7 July the programmes of the Section Sessions (*B6*). These listings were normally in the past tense, but for this session the programme was partly written in the future tense, which suggests that a sub-editor had

---

[19] The manuscripts of some, but by no means all, of the published letters have been located.

overlooked the passage, implying that it may have been taken directly from a printed announcement. Thus, according to the *Literary Gazette*, the speakers on the Saturday morning would be George Kinahan (1829–1908), John Westwood (1805–93), Robert MacAndrew (1802–73), Francis Morris (1810–93) and Draper. The *Literary Gazette* noted that 'The latter three papers will be read about twelve o'clock. Sub-section D [that is, physiology] will adjourn at this hour in order to take part in the discussion' (*B6*). There was no mention of Wilberforce speaking or indeed Huxley, but the adjournment of the sub-section clearly signalled that there would be a discussion of Darwin's work. It would thus appear probable that the section committees arranged for this to happen and probably also at short notice following a disagreement between Huxley and Richard Owen (1804–92) two days earlier over the status of the brain of the gorilla.[20]

That the discussion was hastily arranged receives further confirmation from Draper's account of how the timing of his paper was changed from its original Monday slot:

> On Saturday morning I found that some good friend, I dont know who had very efficiently excelled himself. The paper was charged for that day and appointed to be read in the Museum which is a beautiful room, the Physiological Section adjourned to hear it 'as a mark of the greatest respect in their power to offer to me'[.] There was a notice put up on the front door saying that the reading would take place at 12 oclock[.] A great many carpenters were employed in getting the library ready and benches seats & other conveniences were being carried up. There was a very great audience, the room being filled perfectly full (*A8*).

This, which in common with all the other contemporary accounts, is written as if the writer's contribution was the most significant of the afternoon, provides further evidence for the planning that was put into the discussion. Although no direct evidence has been found, such a reorganisation of the programme together with the elaborate practical arrangements for the discussion could only have been undertaken with the sanction and active participation of the president and secretary of the section, Henslow and Lankester respectively.

By arranging the discussion at short notice, individuals could absent themselves if they so wished without the implication that they had deliberately avoided the issue. Indeed, the list of people who were not present but could have been is nearly as interesting as those who were there. Darwin, of course, did not attend the association; he was pleading illness, as usual, and was in Richmond convalescing.[21] However, a number of people who were at the association did not attend this discussion.

[20] Reported in *Athenaeum*, 7 July 1860, 26. Reprinted in Burkhardt, Smith and Porter, eds, (1985–), **8**, 592–3.

[21] Darwin to Hooker, 2 July 1860, Burkhardt, Smith and Porter, eds, (1985–), **8**, 272–3.

These included the Sandemanian Michael Faraday (1791–1867) (who after being photographed that morning at the top of Christ Church by Charles Lutwidge Dodgson (1832–98),[22] returned to London with a severe headache),[23] the philosopher and historian William Whewell (1794–1866) (*C4*), one of the founders of the Free Church of Scotland, the natural philosopher David Brewster (1781–1868) (*A8*), though his wife was ill and apparently fainted (*A3*) and no evidence has been found that Faraday's colleague at the Royal Institution, Tyndall, was present.[24] Charles Lyell (1797–1875) and Owen also absented themselves,[25] even though the latter had been staying with Wilberforce at Cuddesdon and indeed had given advice for an unfavourable review which Wilberforce had written on the *Origin* for the *Quarterly Review* which was published in July.[26] There is also some evidence which suggests that Huxley intended to leave Oxford on the Saturday afternoon. He noted in his pocket diary the times of the 4.05 and 4.43 trains to Reading,[27] where his family were staying. In making these notes he ignored the six trains before and two after those times.[28] Huxley stayed, however, and the scene was set for the discussion.

The morning session had been held in the library instead of the lecture room normally allocated to the section. Morris, who had been scheduled to speak, was not present and only extracts of his paper were read by Charles Babington (*A11*). The joint afternoon session began with announcements by Charles Daubeny (1795–1867), MacAndrew and Lankester (*B13*). There is no indication that Collingwood's paper drew a large audience; but the account in the *Athenaeum* is clear that it was the announcement of Draper's paper which attracted an immense audience (*B13*): 700 to 1,000 according to Hooker (*A3*),[29] not less than 400 or 500 according to the *Evening Star* (*B1*). The organisers were thus fully justified in moving the meeting into the museum library.

Draper's paper which lasted a 'full hour' according to Draper (*A8*), and an hour and a half according to J.R. Green (*A5*), was entitled 'On the intellectual development of Europe, Considered with reference to the views of Mr. Darwin and others, that

---

[22] Roger Taylor, Edward Wakeling and Peter C. Bunnell, *Lewis Carroll: Photographer. The Princeton University Library Albums*, Princeton, 2002, 251. Dodgson also photographed a number of other participants in the course of the meeting, including Wilberforce and Tristram (*ibid.*) having earlier photographed Huxley (*ibid.*, 249).

[23] Faraday to Moigno, 10 July 1860, Académie des Sciences MS.

[24] A.S. Eve and C.H. Creasey, *Life and Work of John Tyndall*, London, 1945, 84.

[25] *C2* and *A1* respectively.

[26] [Samuel Wilberforce, 'Review of] *On the Origin of Species, by means of Natural Selection; or the preservation of favoured races in the struggle for life*. By Charles Darwin', *Quart. Rev.* (1860), **108**, 225–64.

[27] IC MS HP 70(3). This entry was first noted in Jensen (1988), 173.

[28] *Oxford University Herald*, 30 June 1860, 15.

[29] Peirce agreed with the upper end of this estimate (*A2*).

the progression of organisms is determined by law'. Draper's presentation did not commend itself to Hooker who wrote to Darwin that the paper 'did not mend my temper; for of all the flatulent stuff and all the self sufficient stuffers – these were the greatest' (*A3*). He did not leave, however, nor is it recorded that anyone else did, because Hooker and the others present had heard that 'Soapy Sam was to answer' (*A3*). The discussion was commenced by Richard Greswell (1800–1881) 'who denied that any parallel could be drawn between the intellectual progress of man and the physical development of the lower animals' (*B13*).[30] The *Athenaeum* then reported that the eminent physician and president of the Royal Society, Benjamin Collins Brodie (1783–1862) stated that 'he could not subscribe to the hypothesis of Mr. Darwin. His primordial germ has not been demonstrated to have existed' (*B13*). However, according to Alfred Newton (1829–1907), Brodie 'declared ... that at present it was impossible to say what was the truth' (*A11*). If, as seems possible, each of the speakers had been asked to write up their contribution for the *Athenaeum*, these divergent accounts suggest that, on reflection, Brodie had shifted his position following the discussion. According to Carpenter, a young clergyman then made so ridiculous an intervention that Henslow silenced him with the full support of the audience (*A10*). He then immediately asked Huxley to speak but he declined saying 'he would reply when there were some arguments to meet' (*A10*).

Wilberforce then rose to great applause according to Carpenter (*A10*). According to the *Evening Star* 'He alluded, in a forcible manner, to the weight of authority which had been brought to bear against it [Darwin's theory] – to such men as Sir B. Brodie and Professor Owen – both of whom, with many others eminent for their scientific attainments, had opposed it' (*B1*). Wilberforce attacked Darwin's theory on two grounds, first that it was not a properly inductive theory and, second, that 'All experiments had failed to show any tendency of one animal to assume the form of the other' (*B13*). In particular Wilberforce was anxious to retain the special place of man and asserted that 'The line between man and the lower animals was distinct' (*B13*). One issue that Wilberforce did not refer to was that of the geological problems, which played an important part in his review of the *Origin*.[31] According to J.R. Green, Wilberforce 'proceeded to act the smasher; the white chokers, who were abundant, cheered lustily, a sort of 'Pitch it into him' cheer, and the smasher got so uproarious as to pitch into Darwin's friends – Darwin being smashed – and especially Professor Huxley' (*A5*).

Wilberforce was concerned, above all else, to show conclusively that humans had not evolved from non-human animals, but were specially created. For if this was not the case, there could have been no Adamic fall, and without that the crucifixion and

---

[30] The *Athenaeum* mis-spelt his name.

[31] [Wilberforce] (1860), 239–45.

the resurrection had nothing to redeem and thus Christianity was pointless.[32] This argument was so strong, indeed so far as I am aware has never been confuted, that Huxley used it himself in an essay 30 years later attacking Christianity.[33] Furthermore, it was precisely Wilberforce's point that Kingsley picked up in his parody: 'what would become of the faith, hope, and charity of immortal millions?'[34] were it found that humans and apes were the same. Thus for Wilberforce all was at stake. This doubtless accounts for him asking Huxley 'whether he would prefer a monkey for his grandfather or his grandmother' as *The Press* reported (*B4*), although Arthur Munby (1828–1910) heard the following day that Wilberforce's reference had been to Huxley's father (*C1*). The reason why Wilberforce was prompted to ask this question was because Huxley earlier at the association 'had somewhat facetiously remarked that they [men of science] had nothing to fear even should it be shown that apes were their ancestors' (*B4*).[35] Such a remark would have confirmed the worst fears of the bishop.

Huxley then spoke. According to the *Evening Star* he referred to Wilberforce as an 'unscientific authority' (*B1*). Writing to Frederick Dyster (*c.*1810–93) a couple of months later, Huxley gave his account of what he had said:

> So when I got up I spoke pretty much to the effect – that I had listened with great attention to the Lord Bishops speech but had been unable to discover either a new fact or a new argument in it – except indeed the question raised as to my personal predilections in the matter of ancestry – That it would not have occurred to me to bring forward such a topic as that for discussion myself, but that I was quite ready to meet the Right Rev. prelate even on that ground. If then, said I, the question is put to me would I rather have a miserable ape for a grandfather or a man highly endowed by nature and possessing great means & influence & yet who employs those faculties & that influence for the mere purpose of introducing ridicule into a grave scientific discussion – I unhestiatingly affirm my preference for the ape. Whereupon there was unextinguishable laughter among the people – and they listened to the rest of my argument with the greatest attention. (*A12*)

As with Draper's letter, Huxley placed the best possible interpretation on his actions during the afternoon. According to the *Athenaeum* Huxley 'defended Mr. Darwin's theory from the charge of its being merely an hypothesis. He said, it was an explanation of phenomena in Natural History, as the undulatory theory was for the phenomena of

---

[32] *Ibid.*, 258 where he made this point explicitly.

[33] T.H. Huxley, 'The Lights of the Church and the Light of Science', *Nineteenth Cent.* (1890), **28**, 5–22, 21–2 quoting 1 Corinthians 15: 21–2.

[34] Charles Kingsley, *The Water Babies: a Fairy tale for a land-baby*, London, 1863, 156. For a general discussion of Kingsley's text in the context of discussions on evolution see Nicolaas A. Rupke, *Richard Owen: Victorian naturalist*, New Haven, 1994, 300–303.

[35] For a discussion of this aspect see Jensen (1988), 164.

light. No one objected to that theory because an undulation of light had never been arrested and measured' (*B13*). This statement was not quite accurate, since a number of natural philosophers, such as Brewster, objected to the theory quite strongly, while Faraday had expressed reservations over the years.[36] Huxley continued 'with regard to the psychological distinctions between man and animals; man himself was once a monad – a mere atom, and nobody could say at what moment in the history of his development he became consciously intelligent. The question was not so much one of a transmutation or transition of species, as of the production of forms which became permanent' (*B13*). Thus according to the *Evening Star*, Huxley defended 'the Darwinian theory in an argumentative speech which was loudly applauded' (*B1*).

Argumentative Huxley certainly was, but the following three contributors remained sceptical to one degree or another. Robert Fitzroy (1805–65), the captain of HMS *Beagle* in which Darwin had sailed round the world in the 1830s, said directly after Huxley 'that he regretted the publication of Mr. Darwin's book, and denied Prof Huxley's statement, that it was a logical arrangement of facts' (*B13*). Lionel Smith Beale (1828–1906) 'pointed out some of the difficulties with which the Darwinian theory had to deal, more especially those vital tendencies of allied species which seemed independent of all external agents' (*B13*). John Lubbock (1834–1913) 'emphasised his willingness to accept the Darwinian hypothesis in the absence of any better' (*B13*); faint praise indeed, but Newton reported him as taking 'a decided Darwinian view' (*A11*) at the discussion, again suggesting that the participants shifted position quite quickly after the discussion.[37]

Hooker in his letter to Darwin a couple of days later, did not give Huxley credit for the total success that Huxley claimed. Hooker, who was doubtless also celebrating his 43rd birthday that day, admitted that 'Huxley answered admirably & turned the tables, but he could not throw his voice over so large an assembly, nor command the audience; & he did not allude to Sam's weak points nor put the matter in a form or way that carried the audience' (*A3*). Thus, after Lubbock's intervention, Hooker, no doubt because he thought that Huxley had not made the best of his case, spoke:

> my blood boiled, I felt myself a dastard; now I saw my advantage – I swore to myself I would smite Amalekite Sam hip & thigh if my heart jumped out of my mouth & I handed my name up to the President (Henslow) [Hooker's father-in-law] as ready to thrown down the gauntlet … it … became necessary for each speaker to mount the platform & so there I was cocked up with Sam at my right elbow, & there & then I

[36] Frank A.J.L. James, '"The Optical Mode of Investigation": Light and matter in Faraday's natural philosophy', in David Gooding and Frank A.J.L. James, eds, *Faraday Rediscovered: Essays on the life and work of Michael Faraday, 1791–1867*, London, 1985, 137–61.

[37] This is also implied in a letter George Busk (1807–86) wrote to Lubbock questioning the *Athenaeum*'s account of his contribution. Busk to Lubbock, 16 July 1860, British Library add. MS 49639, f.11.

> smashed him amid rounds of ap[p]lause – I hit him in the wind at the first shot in 10 words taken from his own ugly mouth – & then proceeded to demonstrate in a few more 1 that he could never have read your book & 2 that he was absolutely ignorant of the rudiments of Bot. Science – I said a few more on the subject of my own experience, conversion & wound up with a very few observations on the relative position of the old & new hypotheses, & with some words of caution to the audience – Sam was shut up – had not one word to say in reply & the meeting was dissolved forthwith leaving you [Darwin] master of the field after 4 hours battle (*A3*).

This view of the outcome of the discussion was shared by Huxley who wrote of Wilberforce:

> If he had dealt with the subject fairly & worthily I would not have treated him in this way. But the round mouthed, oily, special pleading of the man ignorant of the subject presumed on his position & his lawyer faculty gave me a most unmitigated contempt for him. You can't think how pleased all his compieres were[;] I believe I was the most popular man in Oxford for full four & twenty hours afterwards (*A12*).

J.R. Green too thought that Huxley had had the best of it: Huxley 'gave his lordship such a smashing as he may meditate on with profit over his port at Cuddesdon' (*A5*).

On the other hand Wilberforce wrote three days later: 'had quite a long fight with Huxley. I think I thoroughly beat him' (*A4*), while Balfour Stewart (1828–87) reported to James David Forbes (1809–68) that 'I think the Bishop had the best of it' (*A7*). Of the newspapers most made no judgement, although *John Bull* suggested the honours should go to Wilberforce (*B7*) as did the High Church *Guardian* which took a sarcastic line against the unnamed Huxley (*B2*).

The certainty with which the supporters claimed victory against the qualified 'I think' claims of their opponents suggests that Huxley may indeed have had the best of the day. The support for Huxley by the audience is perhaps best read as a dislike of Oxford's Bishop.[38] In this context it is significant that the university appointed the Broad churchman Frederick Temple (1821–1902, the first author in *Essays and Reviews* published later in the year)[39] to preach the British Association sermon the following day.[40] Such an appointment by the university can perhaps be best interpreted as a rejection of what Wilberforce stood for in church terms and Huxley gave the dons of the university a focal point to air their views about their bishop. This interpretation would

---

[38] It has been suggested that Wilberforce's appointment to Oxford, by Robert Peel (1788–1850), was to keep the Tractarians in order. Frank M. Turner, *John Henry Newman: the Challenge to Evangelical religion*, New Haven, 2002, 553–4.

[39] Ieuan Ellis, *Seven against Christ: a Study of 'Essays and Reviews'*, Leiden, 1980.

[40] Frederick Temple, *The Present Relations of Science to Religion. A sermon*, Oxford, 1860.

agree with Lyell's comment, written after his return to London: 'Many blame Huxley for his irreverent freedom; but still none of those I heard talk of it, and among them Falconer, assures me the Vice-Chancellor Jeune (a liberal) declared that the Bishop got no more than he deserved' (*C2*). What was thought of the scientific arguments thus becomes immaterial, although unlike most discussions of this character, some opinions were changed in both directions. Wilfrid Simpson (1828–1909) changed from being anti-Darwinian to pro, saying that if what he had heard was all that 'could be said in favour of the old view, he was a convert [to Darwinism]' while Henry Tristram (1829–1906) went in the opposite direction (both in *A11*).[41]

The most interesting newspaper account, and not just because it was the longest, is that which appeared in the *Athenaeum* two weeks later (*B13*). Because of the length one might perhaps expect it to provide the most detailed account and in many ways, of course, it does. But it lacks the passion of the discussion contained in the letters and in the other much shorter newspaper accounts. Gone from this is Wilberforce's jibe and Huxley's response which went to the core of the issues between them. It reads like a bureaucratic account of the discussion and if, as seems likely, the section secretary, Lankester, was responsible for putting this account together, then such a neutering must mean that the Gentlemen of Science of the British Association did not wish to publicise the ungentlemanly behaviour of a bishop and its own savants. This interpretation is supported by there being no reference whatsoever to the discussion in the 1860 *Report of the British Association*. Only a short abstract of Draper's paper was published there.[42] What must have happened is that what seemed a good idea at the time to discuss a new scientific theory, in the course of the passionate debate and ungentlemanly behaviour came to be seen very quickly as something which should be sanitised, if not suppressed altogether. This reading is supported by Munby's comment the following day that 'the proprieties of the [British] Association have been outraged' (*C1*). Furthermore, according to Newton, the discussion was adjourned to the Monday 'but it was then thought by the leaders of both sides that [it] had better be dropped' (*A11*).

Whether it was usual for a section to undertake the composition of an article for the *Athenaeum* is not clear.[43] The effect of the preparation of such a long piece

---

[41] On Tristram's change of mind see I. Bernard Cohen, 'Three Notes on the Reception of Darwin's Ideas on Natural Selection (Henry Baker Tristram, Alfred Newton, Samuel Wilberforce)' in David Kohn, ed., *The Darwinian Heritage*, Princeton, 1985, 589–607, especially 597–8.

[42] J.W. Draper, 'On the Intellectual Development of Europe, considered with reference to the views of Mr. Darwin and others, that the Progression of Organisms is determined by Law', *Rep. Brit. Ass.* (1860), part 2, 115–16.

[43] *Jackson's Oxford Journal*, 30 June 1860, 5, col. b noted the poor accommodation for reporters and 'it is owing to this fact that the London press has scarcely had a representative on the present occasion'.

(just over 2,000 words) which would be seen as a, if not the, definitive account, might render unnecessary the publication of other versions, less acceptable to the sensibilities of the Gentlemen of Science of the British Association. If this was the intention, then, with the exception of Fawcett's piece at the end of the year (*B14*), it was entirely successful. Furthermore, this interpretation points up the reason why the discussion happened at the British Association rather than elsewhere. The leadership of the British Association has only ever had the loosest control over the content of the papers which are read to sections which maintained a high degree of autonomy and sometimes *de facto* independence. The British Association did not, indeed does not, have the centralised, largely consensual, programme which, say, both the Royal Institution and the Royal Society have. As Joe Burchfield has pointed out, Tyndall did not present his controversial views on science and religion in his own Royal Institution, but rather at the British Association.[44] On the basis of this sort of argument one can suggest that the British Association was seen as the place where these kinds of ideological battles might take place. But if they were seen by the leadership of the British Association as damaging the good name of the association, then they would take appropriate action, which, I suggest, is what happened immediately following the events of that afternoon.

This interpretation explains why, apart from Kingsley's 1863 parody in *The Water Babies* and Hooker's at the 1866 Nottingham meeting of the British Association,[45] the first time that the discussion is described in any book or article that I have so far located is in the life of Philip Carpenter published in 1880. This was followed by a mention in the second volume of the *Life of Wilberforce*, written by his son Reginald Wilberforce (1838–1914) and published in 1881. This simply stated:

> The Bishop ... made a long and eloquent speech condemning Mr. Darwin's theory as unphilosophical and as founded on fancy, and he denied that any one instance had been produced by Mr. Darwin which showed that the alleged change from one species to another had ever taken place. In the course of this speech, which made a great impression, the Bishop said, that whatever certain people might believe, he would not look at the monkeys in the Zoological as connected with his ancestors, a remark that drew from a certain learned professor the retort, 'I would rather be descended from an ape than a bishop.'[46]

In the errata of volume three, the final sentence was, presumably at Huxley's insistence, amended thus: 'If I had to choose between being descended from an ape or from a

---

[44] J.D. Burchfield, 'John Tyndall at the Royal Institution', in Frank A.J.L. James, ed., *'The Common Purposes of Life': Science and society at the Royal Institution of Great Britain*, Aldershot, 2002, 147–68, 165.
[45] Huxley (1918), **2**, 102–103.
[46] R.G. Wilberforce, *Life of ... Samuel Wilberforce*, 3 vols, London, 1880–82, **2**, 450–51.

man who would use his great powers of rhetoric to crush an argument, I should prefer the former'.[47]

In 1887, volume two of Darwin's life, edited by his son Francis Darwin (1848–1925), was published. As Table 12.5 shows he went to a good deal of trouble to obtain recollections of the discussion. In addition to quoting from Fawcett's article (*B14*) he contacted Hooker for an account having found his letter to Darwin describing the discussion (*A3*). Hooker did not want Francis to publish this letter on the grounds that it was 'far too much of a *braggart* epistle' (*E2*). However, he agreed to provide another account once he had read the *Athenaeum* description (*B13*).[48] It was not only the *Athenaeum* account that Hooker consulted, for in the 'screed' (*E4*) that he sent Francis for anonymous publication (*E3*),[49] he quoted in turn extracts from Lyell's and Green's letters (*C2*, *A5*). Hooker noted that: 'There was a crowded conversazione in the evening ... where the almost sole topic was the battle of the "Origin," and I was much struck with the fair and unprejudiced way in which the black coats and white cravats of Oxford discussed the question, and the frankness with which they offered their congratulations to the winners in the combat' (*E3*). This, albeit much later recollection, again supports the interpretation of the outcome in terms of Oxford University's dislike of Wilberforce.

Volume two of Darwin's life also contained a chapter by Huxley 'On the Reception of the Origin of Species' in which he severely criticised Wilberforce's review.[50] Reginald Wilberforce rose to defend his father in a letter to *The Times* at the end of which he commented 'Did the lash of Bishop Wilberforce's eloquence sting so sharply that though 27 years have passed, the recollection of the castigation then received is as fresh as ever?' (*E7*) to which Huxley responded 'that an effectual castigation was received by somebody' (*E8*).

The three accounts published during the 1880s (in the lives and letters of Wilberforce, Lyell and Darwin) were clearly sufficient to stir the memory and some of those present went into print. Indeed, Hooker continued to pursue this topic and in March 1887 sent Francis further accounts although these were not published until much later in Hooker's own life (*E5*, *E6*). In 1888 Newton published his recollections of the discussion, which he called 'ever-memorable',[51] but added nothing new.

Francis Darwin must have been dissatisfied with his original account of the discussion, for in preparing his single-volume life of his father (which generally cut

---

47 *Ibid.*, **3**, facing 1.

48 Joseph Dalton Hooker to Francis Darwin, 15 [November?] 1886, CUL MS DAR 119.3, 15.

49 The author was not identified until Huxley (1918), **2**, 303.

50 F. Darwin, *The Life and Letters of Charles Darwin*, 3 vols, London, 1887, **2**, 179–204.

51 A. Newton, 'Early Days of Darwinism', *Macmillan's Magazine* (1888), 58, 241–9, 248–9.

things out), published in 1892, he collected further accounts.[52] These included a description of the discussion (*E9*) which seems to have been sent voluntarily (*E10*) by William Henry Freemantle (1831–1916) following publication of volume two of Darwin's life. Francis also published Green's letter (*A5*) and a letter from Huxley to him commenting on Freemantle's letter and recollecting how he had come to be present:

> The odd part of the business is, that I should not have been present except for Robert Chambers. I had heard of the Bishop's intention to utilise the occasion. I knew he had the reputation of being a first-class controversialist, and I was quite aware that if he played his cards properly, we should have little chance, with such an audience, of making an efficient defence. Moreover, I was very tired, and wanted to join my wife at her brother-in-law's country house near Reading, on the Saturday. On the Friday I met Chambers in the street, and in reply to some remark of his, about his going to the meeting, I said that I did not mean to attend it – did not see the good of giving up peace and quietness to be episcopally pounded. Chambers broke out into vehement remonstrances and talked about my deserting them. So I said, 'Oh! if you are going to take it that way, I'll come and have my share of what is going on.'
>
> So I came, and chanced to sit near old Sir Benjamin Brodie. The Bishop began his speech, and to my astonishment very soon showed that he was so ignorant that he did not know how to manage his own case. My spirits rose proportionately, and when he turned to me to with his insolent question, I said to Sir Benjamin, in an undertone, 'The Lord hath delivered him into mine hands.'
>
> That sagacious old gentleman stared at me as if I had lost my senses. But, in fact, the Bishop had justified the severest retort I could devise, and I made up my mind to let him have it. I was careful, however, not to rise to reply until the meeting called for me – then I let myself go. (*E11*)[53]

In the late 1890s Isabel Sidgwick published another eye-witness account. She too called it 'memorable'.[54] She recollected that Wilberforce had said that 'rock-pigeons had been what rock-pigeons had always been',[55] which was one of the points he had made in his *Quarterly Review* article.[56] Like Hooker's anonymous piece in the life of Darwin, Sidgwick commented on the evening at Daubeny's where 'every one was eager to congratulate the hero of the day. I remember that some naïve person wished "it could come over again;" and Mr. Huxley, with the look on his face of the victor

---

[52] Which he seems to have continued after publication with letters from Stoney (*E12* and *E13*). These recollected both Fitzroy's presence and the grandparent incident. As he placed the event in the Sheldonian Theatre, Stoney's accuracy may be doubted.

[53] The reference to meeting Chambers on the Friday confirms that there was some degree of preparation for the meeting, while Huxley's reference to going to Reading explains the train times in his pocket diary.

[54] [I. Sidgwick], 'A Grandmother's Tales', Macmillan's Magazine, 1898, **78**, 433.

[55] *Ibid*., 433.

[56] [Wilberforce] (1860), 234–5.

who feels the cost of victory, put us aside saying, "Once in a life-time is enough, if not too much."'[57]

In 1900 the life and letters of Huxley appeared, edited by his son Leonard. This contained the longest account to date of the discussion. Leonard wanted to make the point that the discussion had been a, if not the, crucial event in Huxley's life. He commenced, by promoting the discussion from memorable to 'The famous Oxford Meeting of 1860 was of no small importance in Huxley's career. It was not merely that he helped to save a great cause from being stifled under misrepresentation and ridicule – that he helped to extort for it a fair hearing; it was now that he first made himself known in popular estimation as a dangerous adversary in debate – a personal force in the world of science which could not be neglected'.[58] However, despite the promotion from memorable to famous, Leonard had a problem with the scarcity of evidence. Indeed he did not even have Huxley's letter to Dyster (*A12*). He used the *Athenaeum* account (*B13*) and the lives and letters of Lyell (*D3*) and Darwin (*D6, D8*) as well as Sidgwick's piece (*D11*). But this was not enough for him. Because he believed it to be a turning point in his father's career, he wanted as much information on it as possible and therefore contacted other eye-witnesses to provide reports of a discussion nearly 40 years earlier. Adam Farrar (1826–1905) gave him an account (*E17*), that disagreed with Sidgwick about what was said.[59] Augustus George Vernon Harcourt (1834–1919) elaborated on the discussion (*E16*) and, finally, Leonard reprinted from the life of Darwin, Freemantle's account (*E9*) and Huxley's letter to Francis Darwin (*E11*).

As I noted earlier, the point of Leonard's narrative was to demonstrate that the discussion was an 'open clash between Science and the Church'[60] and that the 'importance of the Oxford meeting lay in the open resistance that was made to authority'.[61] It is this universal interpretation of the discussion that has entered into modern consciousness following the concerns of the late nineteenth century. Gone from this story is the local context of the unpopularity of a local bishop which helped determined the outcome. Had Wilberforce not been so unpopular in Oxford he would have carried the day and not Huxley. It was the audience who by their taking the opportunity of changing sides, allowed Huxley to win and enabled the discussion to be used as strong evidence for the antagonism between science and religion. Aside from

---

[57] [Sidgwick] (1898), 434.

[58] Huxley (1900), **1**, 179.

[59] Although this letter admitted that Huxley had secured a victory, Farrar sharply defended Wilberforce; Leonard omitted those parts from the text that he published. Francis Darwin expressed 'a perfectly unfounded prejudice against Canon Farrar'. Francis Darwin to Leonard Huxley, 3 December 1896, IC MS HP 13, f.80–81.

[60] Huxley (1900), **1**: 181.

[61] *Ibid.*, **1**: 189.

Reginald Wilberforce's brief intervention (*E7*), there were no accounts written by the other side to interpret the discussion in terms favourable to themselves. By turning this local discussion into a universal myth, men and women in the late nineteenth century contributed to the process of separating in some contexts science from Christian belief and made a significant contribution to the separation of the professional cultures from each other.[62] These separations were thus a construction of the specific interests of the protagonists who, using very limited evidence from a discussion decades earlier, were able to construct a myth about the relations of science and religious belief that has become every bit as powerful as the dispute between Galileo and the Church in the early seventeenth century.[63] Such myths, whatever their historical shortcomings, do perform a service by allowing modern practitioners to discourse generally; but this tells us far more about their own views about the relations of science and religion, about the separation or otherwise of the cultures, than it does about a four-hour discussion that happened in the summer of 1860.

Historians tend, either consciously or unconsciously, to attach importance to the events they study. In this case I have been seeking deliberately to minimise the importance of what I have been writing about. As should have become clear by now, my view is that the significance of the discussion lies in its historiographical consequences. However, myths, being a means of communication, are notoriously hard to shift, even with careful historical research. They are too useful and too well-entrenched to be countered effectively, at least in the short term. My own Waterloo came when a much earlier version of this paper was read to the Birmingham meeting of the British Association in 1996 and was published in the *Birmingham Post*.[64] Writers in newspapers have no control over the headline under which their work appears. These are written by sub-editors who, in this case, seemed to be unable to understand what they read, and thus the headline above my article read 'Science v Religion: The Big Match'.

---

[62] Frank M. Turner, 'The Victorian Conflict between Science and Religion: a Professional dimension', *Isis* (1978), **69**, 356–76.

[63] Brooke (1991), 99–109.

[64] *The Birmingham Post*, 21 October 1996, supplement, 24.

Table 12.1 Contemporary Diaries and Letters

A1	Samuel Wilberforce Diary	30 June 1860[1]
A2	Benjamin Peirce Diary	30 June 1860[2]
A3	Joseph Dalton Hooker to Charles Darwin	2 July 1860[3]
A4	Samuel Wilberforce to Charles Anderson	3 July 1860[4]
A5	John Richard Green to W. Boyd Dawkins	3 July 1860[5]
A6	Antonia Draper to Family	3 July 1860[6]
A7	Balfour Stewart to James David Forbes	5 July 1860[7]
A8	John William Draper to Family	6 July 1860[8]
A9	Hugh Falconer to Darwin	9 July 1860[9]
A10	Philip Pearsall Carpenter to Joseph Henry	July 1860[10]
A11	Alfred Newton to Edward Newton	25 July 1860[11]
A12	Thomas Henry Huxley to Frederic Daniel Dyster	9 September 1860[12]

1 Bodleian MS Wilberforce dep. e.327.
2 N. Reingold, *Science in Nineteenth-Century America: a Documentary history*, New York, 1964, 197–8.
3 Burkhardt, et al. (1985–), **8**, 270–72.
4 Bodleian MS Wilberforce d.29, f.30–31.
5 Leslie Stephen, *Letters of John Richard Green*, London, 1901, 142–5.
6 Library of Congress, Draper papers, Box 45.
7 St Andrews University MS JDF 1860/133.
8 Library of Congress, Draper papers, Box 44.
9 Burkhardt, et al. (1985–), **8**: 281–2.
10 R.L. Carpenter, *Memoirs of the Life and Work of Philip Pearsall Carpenter*, London, 1880, 243–9.
11 A.F.R. Wollaston, *Life of Alfred Newton*, London, 1921, 118–20.
12 Imperial College MS HP 15, f.115–18. Partly published in D.J. Foskett, 'Wilberforce and Huxley on Evolution', *Nature* (1953), **172**, 920.

Table 12.2 Contemporary Newspaper and Periodical Reports

B1	*The Evening Star*	2 July 1860, 3, col. b
B2	*The Guardian*	4 July 1860, 593
B3	*The Athenaeum*	7 July 1860, 19[1]
B4	*The Press*	7 July 1860, 656
B5	*The Literary Gazette*	7 July 1860, 807
B6	*The Literary Gazette*	7 July 1860, 812
B7	*John Bull*	7 July 1860, 422
B8	*The Inquirer*	7 July 1860, 566
B9	*Oxford University Herald*	7 July 1860, 8
B10	*Oxford Chronicle*	7 July 1860, 2. col. e
B11	*Jackson's Oxford Journal*	7 July 1860, 2, col. f
B12	*Illustrated London News*	7 July 1860, 3

B13	*The Athenaeum*	14 July 1860, 64–5[2]
B14	Henry Fawcett, 'A Popular Exposition of Mr. Darwin on the Origin of Species'	*Macmillan's Mag.* 1860, 3, 81–92, 88

1 This was reprinted in Burkhardt, et al. (1985–), **8**, 590–91.
2 This was reprinted in *ibid.*, **8**, 593–7.

Table 12.3 Hearsay Reports

C1	Arthur J. Munby Diary	1 July 1860[1]
C2	Charles Lyell to Charles Bunbury	4 July 1860[2]
C3	Mountstuart E. Grant Duff Dairy	4 July 1860[3]
C4	William Whewell to James David Forbes	24 July 1860[4]

1 D. Hudson, *Munby...*, London, 1972, 64–5.
2 K. Lyell, *Life, Letters and Journals of Sir Charles Lyell*, 2 vols, London, 1881, **2**, 334–6.
3 Mountstuart E.G.D., *Notes from a Diary 1851–1872*, 2 vols, London, 1897, **1**, 139.
4 St Andrews University MS JDF 1860/145(a).

Table 12.4 Later Texts Describing the Discussion (in Chronological Order)

D1	Philip Pearsall Carpenter[1]	1880
D2	Samuel Wilberforce[2]	1881
D3	Charles Lyell[3]	1881
D4	George Rolleston[4]	1884
D5	Henry Fawcett[5]	1885
D6	Charles Robert Darwin[6]	1887
D7	Alfred Newton[7]	1888
D8	Charles Robert Darwin[8]	1892
D9	Edward Bagnall Poulton[9]	1896
D10	Charles Cardale Babington[10]	1897
D11	Isabel Sidgwick[11]	1898
D12	Thomas Henry Huxley[12]	1900
D13	William Tuckwell[13]	1900
D14	John Richard Green[14]	1901
D15	Henry Wentworth Acland[15]	1903
D16	James Bryce[16]	1909
D17	Joseph Dalton Hooker[17]	1918
D18	Alfred Newton[18]	1921

1 Carpenter (1880), 243–9.
2 R.G. Wilberforce, *Life of ... Samuel Wilberforce*, 3 vols, London, 1880–82, **2**, 450–51.
3 Lyell (1881), **2**, 334–6.

4 E.B. Tylor, 'Biographical Sketch', in William Turner, ed., *Scientific Papers and Addresses by George Rolleston*, 2 vols, London, 1884, **1**, ix–xv, xxxiii–xxxiv.
5 L. Stephen, *Life of Henry Fawcett*, London, 1885, 99.
6 F. Darwin, *The Life and Letters of Charles Darwin*, 3 vols, London, 1887, **2**, 320–23.
7 A. Newton, 'Early Days of Darwinism', *Macmillan's Magazine* (1888), **58**, 241–9, 248–9.
8 F. Darwin, *Charles Darwin*, London, 1892, 236–43.
9 E.B. Poulton, *Charles Darwin and the Theory of Natural Selection*, London, 1896, 153–6.
10 J.E.B. Mayor, *Memorials ... Charles Cardale Babington*, London, 1897, xx, xxxi.
11 [I. Sidgwick], 'A Grandmother's Tales', *Macmillan's Magazine*, 1898, **78**, 425–35, 433–4. Author identified in W.E. Houghton, ed., *Wellesley Index ... vol. 1*, Toronto, 1966, 656.
12 Huxley (1900), **1**, 179–89.
13 William Tuckwell, *Reminiscences of Oxford*, London, 1900, 50–54.
14 L. Stephen, *Letters of John Richard Green*, London, 1901, 142–5.
15 J.B. Atlay, *Sir Henry Wentworth Acland ... A Memoir*, London, 1903, 302–3.
16 James Bryce, 'Personal Reminiscences of Charles Darwin and the Reception of the "Origin of Species"', *Proceedings of the American Philosophical Society* (1909), **48**, iii–xiv, xi–xii.
17 L. Huxley, *Life ... Sir Joseph Dalton Hooker*, 2 vols, London, 1918, **1**, 520–27 and **2**, 303.
18 A.F.R. Wollaston, *Life of Alfred Newton*, London, 1921, 118–20.

Table 12.5 Later Letters

E1	Julius Victor Carus to Francis Darwin	[*c.*1886][1]
E2	Joseph Dalton Hooker to Francis Darwin	30 October 1886[2]
E3	[Joseph Dalton Hooker] to Francis Darwin	*c.*21 November 1886[3]
E4	Joseph Dalton Hooker to Francis Darwin	21 November 1886[4]
E5	Joseph Dalton Hooker to Francis Darwin	1887[5]
E6	Joseph Dalton Hooker to Francis Darwin	10 March 1887[6]
E7	Reginald Wilberforce to *The Times*	28 November 1887[7]
E8	Thomas Henry Huxley to *The Times*	30 November 1887[8]
E9	William Henry Freemantle to Francis Darwin	[*c.*1888][9]
E10	William Henry Freemantle to Francis Darwin	28 July 1888[10]
E11	Thomas Henry Huxley to Francis Darwin	27 June 1891[11]
E12	George Johnstone Stoney to Francis Darwin	17 May 1895[12]
E13	George Johnstone Stoney to Francis Darwin	18 May 1895[13]
E14	John Lubbock to Francis Darwin	2 January 1896[14]
E15	Isabel Sidgwick to Leonard Huxley	[late 1890s][5]
E16	Augustus George Vernon Harcourt to Leonard Huxley	9 July 1899[16]
E17	Adam Storey Farrar to Leonard Huxley	12 July 1899[17]

1 Darwin (1887), **2**, 322.
2 Huxley (1918), **2**, 303. CUL MS DAR 199.3, 14.
3 Darwin (1887), **2**, 321–3, sent with *E4*.
4 Huxley (1918), **2**, 303. CUL MS DAR 199.3, 16.
5 *Ibid.*, **2**, 303.
6 *Ibid.*, **2**, 304. CUL MS DAR 199.3, 18.

7 *The Times*, 29 November 1887, 10, col. d.
8 *The Times*, 1 December 1887, 8, col. d.
9 Darwin (1892), 238–9.
10 CUL MS DAR 107: 21–2.
11 Darwin (1892), 240–41.
12 CUL MS DAR 107, 36–9
13 *Ibid.*, 40–41.
14 *Ibid.*, 30.
15 Huxley (1900), **1**, 185, 188–9.
16 *Ibid.*, **1**, 185. IC MS HP 18, f.1–2.
17 *Ibid.*, **1**, 182–3, 183–4. IC MS HP 16, f.13–19.

Table 12.6 Unpublished References 1860–80

F1	Kingsley to Huxley	28 February 1862[1]
F2	Daubeny to Darwin	5 July 1862[2]
F3	Carus to Darwin	15 November 1866[3]

1 IC MS HP 19, f.203–4. Alludes in passing to the discussion.
2 Burkhardt, et al. (1985–), **10**, 301–2. Noted that in his recent lecture on orchids 'it was the first time that your views had been publickly noticed at Oxford since the famous discussion in which the Bishop of Oxford & Huxley played so prominent a part'.
3 CUL MS DAR 161, 54. 'I shall never forget that meeting of the combined sections of the British Association when at Oxford 1860.'

Table 11.7 Those Identified as Being Present

1	2	3	4	5	6	7	8
Henry Wentworth Acland	*D15*	1815–1900	45	DNB	Physician	P	Oxford
Charles Cardale Babington	*A11*	1808–95	52	DNB	Botanist	Z	Cambridge
Lionel Smith Beale	*B13*	1828–1906	32	DNB	Physician	P	Oxford
Jane Brewster	*A3*				Brewster's wife		Edinburgh
Benjamin Collins Brodie	*B13*	1783–1862	77	DNB	Surgeon, PRS	P	London
Benjamin Collins Brodie	*E17*	1817–80	43	DNB	Chemist		Oxford
James Bryce	*D16*	1838–1922	22	DNB	Classicist		Oxford
Philip Pearsall Carpenter	*A10*	1819–77	21	DNB	Presbyterian minister		Warrington
Julius Victor Carus	*F3*	1823–1903	37	DSB	Zoologist	PZ	Leipzig
Robert Chambers	*E11*	1802–71	58	DNB	*Vestiges* author		Edinburgh
Cuthbert Collingwood	*A11*	1826–1908	34	DNB	Naturalist	Z	Oxford
Charles Giles Bridle Daubeny	*B13*	1795–1867	65	DNB	Chemist	Z	Oxford
Joseph Dayman	*A3*	d. *c.*1868			Naval officer		
John Dingle	*E17*	1812–*c.*86	48	AC	Clergyman		Durham
Antonia Gardner Draper	*A6*	*c.*1814–70	46		Draper's wife		New York
John William Draper	*A3*	1811–82	49	DNB	Chemist	Z	New York
Virginia Draper	*A6*				Draper's daughter		New York
Thomas Simpson Evans	*E17*	1797–80	63	AC	Clergyman		London
Hugh Falconer	*C2*	1808–65	52	DNB	Palaeontologist		London
Adam Storey Farrar	*E17*	1826–1905	34	DNB	Theologian		Oxford
Henry Fawcett	*B14*	1833–84	27	DNB	Blind politician		Cambridge
Robert Fitzroy	*B13*	1805–65	55	DNB	Scientific administrator		London
Michael Foster	*D12*	1836–1907	24	DNB	Physiologist	P	London
William Henry Freemantle	*E9*	1831–1916	29	WWW	Clergyman		Oxford
John Richard Green	*A5*	1837–83	23	DNB	Historian		Oxford
Thomas Hill Green	*E17*	1836–82	24	DNB	Philosopher		Oxford
Richard Greswell	*B13*	1800–1881	60	DNB	Tutor		Oxford
Augustus George Vernon Harcourt	*E16*	1834–1919	26	DNB	Chemist		Oxford

Table 11.7 (continued)

1	2	3	4	5	6	7	8
John Stevens Henslow	*A3*	1796–1861	64	DNB	Botanist	Z	Cambridge
Joseph Dalton Hooker	*A3*	1817–1911	43	DNB	Botanist	Z	Kew
William Jackson Hooker	*E16*	1785–65	75	DNB	Botanist		Kew
Thomas Henry Huxley	*A3*	1825–95	35	DNB	Naturalist	ZP	London
Jenkins	*A5*						
Francis Jeune	*C2*	1806–68	54	DNB	Vice-chancellor		Oxford
George Henry Kinahan	*B6*	1829–1908	31	DNB	Geologist		Dublin
Edwin Lankester	*B13*	1814–74	56	DNB	Naturalist	Z	London
Humphry Lloyd	*E12*	1800–1881	60	DNB	Natural Philosopher		Dublin
John Lubbock	*B13*	1834–1913	26	DNB	Naturalist	ZP	London
Robert MacAndrew	*B13*	1802–73	58	*	Naturalist	Z	Liverpool
Richard Monkton Milnes	*C3*	1809–85	51	DNB	Politician		London
Alfred Newton	*A11*	1829–1907	31	DNB	Zoologist	Z	Cambridge
Benjamin Peirce	*A2*	1809–80	51	DAB	Mathematician		Boston
George Rolleston	*D4*	1829–81	31	DNB	Physician		Oxford
Isabel Sidgwick	*D11*						
William Sidgwick	*D9*						
Wilfrid Huddleson Simpson	*A11*	1828–1909	32	DNB	Geologist		Cambridge
Balfour Stewart	*A7*	1828–87	32	DNB	Physicist		Kew
George Johnstone Stoney	*E12*	1826–1911	44	DNB	Natural Philosopher		Dublin
Henry Baker Tristram	*A11*	1829–1906	31	DNB	Naturalist	Z	Oxford
William Tuckwell	*D13*	1829–1919	31	WWW	Schoolmaster		Oxford
John Obadiah Westwood	*B6*	1805–93	55	DNB	Zoologist	Z	Oxford
Samuel Wilberforce	*A3*	1805–73	54	DNB	Bishop		Oxford

Column 1: Name; **2:** Evidence of presence at discussion; **3:** Vital dates; **4:** Age in 1860 (average 43.3); **5:** Source of biographical information (DNB = *Dictionary of National Biography*; AC = *Alumni Cantabrigienses*, * = Burkhardt, et al. (1985–), **9:** 539); **6:** Profession; **7**: Membership of Zoological (Z) or Physiological (P) Committee; **8:** Geographical location.

Chapter 13

The Invention of Altruism: Auguste Comte's *Positive Polity* and Respectable Unbelief in Victorian Britain

Thomas Dixon[1]

Introduction: The Language of 'Altruism'

At 10.40pm, on Tuesday March 27th 1883, at Windsor Castle, one of the royal household died. This was not the death of just any royal domestic, but of the man who had for over 18 years been Her Majesty's faithful personal attendant, and, according to some rumours, much more than a personal attendant. This was, of course, John Brown. On the following Sunday, the preacher at the Queen's Private Chapel at Windsor was the Reverend William Boyd Carpenter, the Bishop of Ripon and chaplain to the Queen. He took as the subject for his sermon, later to be published at the Queen's command, 'Christian Altruism'. Carpenter told the grieving Queen, her family and attendants, that human social life was one of 'absolute interdependence', and that the great truth taught by nature was 'the solidarity of all life'.[2] Carpenter's sermon made use of both scientific and religious ideas; he looked to both nature and scripture as sources of moral authority. He spoke of the discoveries of Newton and Galileo as not inventions but unveilings, unveilings which revealed pre-existing natural laws. Similarly, he went on, St Paul had unveiled the pre-existing moral law that 'No man liveth to himself'.[3] Christ himself had told his disciples that 'The greatest is he that doth serve'.[4] The

---

[1] For institutional and financial support for the research from which this essay arises, I am grateful to the British Academy for a Postdoctoral Research Fellowship, to Churchill College for a Junior Research Fellowship, and to the Cambridge Faculty of Divinity. I am grateful to Léon Turner, Fraser Watts, and Emily Butterworth for reading and commenting on an earlier version of this essay, and to participants in seminars at which I have presented versions of this material at Canterbury Christ Church University College, Heythrop College, London, and the British Academy Postdoctoral Fellowship Symposium for their very valuable comments and questions.

[2] William Boyd Carpenter, *Christian Altruism: a Sermon preached in the Private Chapel, Windsor Castle, Sunday 1 April, 1883*, London, 1883, 5.

[3] Romans 14:7, quoted in *ibid.*, 6.

[4] Luke 22:26, quoted in *ibid.*, 9.

virtuousness of service was indeed an appropriate theme in the circumstances, and one which Carpenter warmed to: 'men may win a glory, than which even those who occupy the higher departments can win no higher, the glory of having done their duty of simple and honest service'. John Brown's 'continuous and unselfish service' had bestowed upon his life a dignity to which all might aspire.[5] So Carpenter's sermon elevated John Brown alongside Christ as an example of true altruism, a living sacrifice labouring and suffering always for others rather than for self. The Queen must have been moved to hear such a eulogy for the man who had devoted himself to her service. Later that year she would erect a statue in Brown's honour at Balmoral, inscribed with the following lines, penned by Tennyson at her request: 'Friend more than servant, loyal, truthful, brave! Self less than duty even to the grave!'[6]

This talk of sacrificing oneself and living for others, and the very term 'altruism', however, were not simply parts of a traditional Christian idiom. Rather, they had come into the language of British philosophers and theologians through the influence of a controversial French atheist, the philosopher and sociologist Auguste Comte (1798–1857). He had coined the term '*altruisme*' around 1830, and used it in a published work for the first time in 1851.[7] It had been imported into English usage by his British admirers. How, then, by the 1880s, had 'altruism' come to be adopted by an Anglican bishop as the expression of a conservative and Christian ideal of service within the British social hierarchy? To whom did the new Comtean terminology appeal, and why? Did the use of the language of 'altruism' indicate familiarity with the latest scientific ways of thinking? Did it indicate a commitment to a particular kind of religion? I hope to suggest some answers to these questions by looking in a little more detail at the story of the development of the language of 'altruism' in Britain in the second half of the nineteenth century, and by reflecting on the different religious and political projects that were at stake as this discourse developed.[8]

This story is instructive in several different respects. First, it provides a new perspective on the well-worn topic of relationships between religion and science in Victorian Britain. There have been many different ways of telling stories about these

---

[5] *Ibid.*, 10–11.

[6] Quoted in Hope Dyson and Charles Tennyson, eds, *Dear and Honoured Lady: the Correspondence between Queen Victoria and Alfred Tennyson*, London, 1969, 107.

[7] Some authorities credit Comte's teacher, François Andrieux with the original coinage; see Maxime Leroy, *Histoire Des Idées Sociales En France*, 3 vols, Paris, 1946–54, iii, 102–3; Centre National de la Recherche Scientifique, *Trésor de la Langue Française: Dictionnaire de la Langue du XIXe et du XXe Siècle , 1789–1960*, 16 vols, Paris, 1971–94, ii, 638–9. The first published usage seems to be in the first volume of Comte's *Système de Politique Positive*, 4 vols, Paris, 1851–54.

[8] This essay is a brief and preliminary attempt to tell this story, which I hope to develop in more depth and detail in future work.

engagements, often focusing on debates about Darwin, geology, evolution, biblical criticism, scientific professionalisation, or religious tests at Oxford and Cambridge. Although developments in the natural sciences certainly were the occasion for much religious and theological debate, they were not the only scientific developments causing such debate. Attempts made by Auguste Comte and others to develop scientific accounts of ethics and of society constituted as widely discussed a threat to religious orthodoxy in Britain for most of the Victorian period as any geological or evolutionary theories, even those associated with the name of Charles Darwin.

Secondly, it is illuminating to compare debates about science, religion and altruism in the present day with these Victorian discussions. In recent decades much scholarly discussion of altruism has been focused around the reception and development of sociobiological theories first put forward in the 1970s by writers such as E.O. Wilson and Richard Dawkins.[9] Dawkins' 'selfish gene' theory, in particular, has inspired theologians to get to grips with evolutionary and psychological theories of altruism, especially those theologians who take altruism to be the defining characteristic of religious virtue.[10] Dawkins' book *The Selfish Gene*, first published in 1976, is still the single work that casts the longest shadow over this area of debate. Dawkins' argument was that evolution should be thought of as a process favouring the reproduction not primarily of well-adapted individuals but of well-adapted genes. In his picture, the individual was more or less at the mercy of his or her genes. However, there was some hope – we could fight against the tyranny of these 'selfish replicators'.[11] 'Let us try to *teach* generosity and altruism', Dawkins wrote, 'because we are born selfish. Let us understand what our selfish genes are up to, because we may then at least have the chance to upset their designs, something that no other species has ever aspired to do.'[12] Commentators have noted that recent popular science books often have, as in this quotation from Dawkins, a zealous and pious air about them. However, the faith being preached is not a traditional theistic one. It is, rather, a combination of humanistic moralising and scientific atheism.[13]

It is interesting to find that contemporary writing about altruism combines scientific and religious elements in this way, since, as I show below, the science of altruism was religious from the outset. There are differences too, though. In recent debates, it has

---

[9] Richard Dawkins, *The Selfish Gene*, new edn, Oxford and New York, 1989; Edward O. Wilson, *Sociobiology: the New synthesis*, Cambridge, Mass., 1975.

[10] See, for instance, Colin Grant, *Altruism and Christian Ethics*, Cambridge, 2001; Stephen G. Post, et al., eds, *Altruism and Altruistic Love: Science, philosophy and religion in dialogue*, Oxford and New York, 2002.

[11] Dawkins, 1989, 201.

[12] *Ibid.*, 3.

[13] See Thomas Dixon, 'Scientific Atheism as a Faith Tradition', *Studies in History and Philosophy of Biological and Biomedical Sciences* (2002), **33**, 337–59.

often been the scientists who have denied the existence of 'genuine' altruism, and the theologians who have contradicted this and insisted on the innate goodness of human motivations. When we cast our eyes back to nineteenth-century uses and contexts of the term 'altruism', a rather different picture emerges. Both the scientific and the religious contexts were different – instead of genetics and evolution, the contested sciences of sociology and phrenology were at the centre of the picture. Instead of tacit humanistic religiosity, we find that engagements between Christian faith and explicitly religious unbelief and humanism were driving forces. And instead of scientific thinkers denying that genuine altruism was found in nature and theologians defending its innateness, the positions on this question seem more often to have been reversed. As I shall show, different contexts produced different meanings.

The growth and development of discourses of 'altruism' in Victorian Britain illustrates not only the impact of Comtean thought and language, but also something important specifically about attitudes to religion and theology. While there was much anti-theological feeling amongst liberal writers, many of whom had a high regard for the sciences, many still felt the pull of a strong religious impulse. Could the new ethical discourse of 'altruism' provide such individuals with a way to articulate atheistic and anti-theological views in a way that did not give up on ethics and religion altogether? 'Altruism' was first used in a published work in English just one year after its first published use by Comte. This was in an 1852 review of the *Positive Polity* by the English philosopher (better known as the partner of George Eliot) G.H. Lewes, who described it as 'a felicitous phrase coined by Comte'.[14] Over the next 20 years the word 'altruism' established itself more securely in the world of Victorian moral philosophy, through its use by writers with sympathies with Comte and his 'Positive philosophy': especially Lewes, George Eliot, John Stuart Mill, Herbert Spencer; the philosopher and surgeon James Hinton, who introduced the term into the discussions of the Metaphysical Society; and Hinton's son-in-law, the literary critic John T. Nettleship.[15] Certainly most of these early users of the term were writers alienated from Christian theology and more or less inclined towards some form of scientific agnosticism and secularism. However, from an early stage, there was a great interest among religious and

---

[14] [G.H. Lewes], 'Contemporary Literature of France', *Westminster Review* (1852), **58**, 617–18.

[15] Alan W. Brown, *The Metaphysical Society: Victorian minds in crisis 1869–1880*, New York, 1947, 124; [George Eliot], 'The Natural History of German Life', *Westminster Review* (1856), **66**, 55; Ellice Hopkins, ed., *Life and Letters of James Hinton*, London, 1878, 194; [Lewes] (1852), **58**, 617–18; *idem*, *Comte's Philosophy of the Sciences*, London, 1853, 217; John Stuart Mill, 'Later Speculations of Auguste Comte', *Westminster Review* (1865), **28**, 1–42; John T. Nettleship, *Essays on Robert Browning's Poetry*, London, 1868, ch. 6, 167; Herbert Spencer, *The Data of Ethics*, London, 1879, chs 12 and 13; *idem*, 'The Study of Sociology. VII. Subjective Difficulties – Emotional', *Contemporary Review* (1873), **21**, 159–82.

theological thinkers too. This religious dimension to Victorian discussions of altruism is one that has until now been somewhat overlooked. In his *Public Moralists*, Stefan Collini provides a very rich study of the 'culture of altruism' in Britain, especially as it flourished between the 1850s and 1880s.[16] Collini correctly identifies this culture as one of great moral seriousness, in which obligation and duty were constant themes; he also charts, to some extent, the way that calls for absolute 'altruism' developed from their original (and most extreme) form in Comte into the wider culture; and how this ideal particularly appealed to those who sought religious fulfilment outside traditional Christianity. What Collini's account does not quite do justice to, however, focusing as it does on two primarily secular thinkers (John Stuart Mill and Leslie Stephen), is the full spectrum of religious sentiments, both Christian and anti-Christian, that infused the discourse of 'altruism'. This term not only featured as the centrepiece of the Comtean Religion of Humanity, as Collini acknowledges, but it could also be identified as itself a specifically Christian virtue (this was just the identification made by Boyd Carpenter in his 1883 sermon about John Brown). So, when Collini selectively quotes from the young Henry Sidgwick, writing in 1861, as follows: 'The strongest conviction I have is a belief in what Comte calls "*altruisme*"', what his readers would not realise is that this section of Sidgwick's journal continues with the words: 'the cardinal doctrine, it seems to me, of Jesus of Nazareth'.[17] I hope to bring out more strongly, then, the tensions and interminglings between godless religiosity and Christianity that Victorian discussions of 'altruism' allowed.

Auguste Comte and the Religion and Science of Altruism

From the appearance of the first volume of Comte's *Cours de Philosophie Positive*, in 1830, onwards, liberal intellectuals in Britain started to engage enthusiastically with this controversial new way of thinking about science and society. John Stuart Mill recommended the *Cours* to his protégé, Alexander Bain, as being, despite certain flaws, 'very nearly the grandest work of this age'.[18] Five more volumes appeared during the following 12 years, and a freely translated English abridgement, produced by the journalist and political writer Harriet Martineau, appeared in 1853.[19] Thomas Huxley, writing in the *Westminster Review*, praised Martineau for converting the

---

[16] Stefan Collini, *Public Moralists: Political thought and intellectual life in Britain 1850–1930*, Oxford, 1991, ch. 2.

[17] *Ibid.*, 86; J.B. Schneewind, *Sidgwick's Ethics and Victorian Moral Philosophy*, Oxford, 1977, 42.

[18] Quoted in Alexander Bain, *Autobiography*, London, 1904, 112.

[19] Harriet Martineau, *The Positive Philosophy of Auguste Comte, Freely Translated and Condensed*, 2 vols, London, 1853.

Cours, 'without the loss of a sentence that was worth keeping, from six wearisome volumes of indifferent French, into two of very excellent readable English'.[20] Even Comte himself seems to have agreed that Martineau had improved upon the original, since he included her abridgement, instead of the *Cours* itself, among the 150 volumes of the 'Positivist Library', which constituted the recommended reading for devotees of positivism.[21]

Through reading Comte's *Cours*, the Martineau abridgement and articles and reviews in the *Westminster Review* and elsewhere, an increasing proportion of the reading public became familiar with new Comtean terms and ideas. Two of the most significant and widely discussed of these were Comte's 'law of the three states' of intellectual development and his hierarchy of the sciences. The law of the three states said that the development of the human mind, and of the various sciences through history, always passed through three states: the theological, the metaphysical and the 'positive' or scientific. The commitment to attaining the positive state in all areas of study – the state in which all supernatural and metaphysical agencies and abstractions would be abandoned in favour of strict and purely empirical laws – was to be the central commitment of the 'positivist' philosopher. The second significant suggestion debated among Comte's British readers was his idea that the sciences could be arranged in a 'hierarchy', ascending from mathematics, through astronomy and physics, to chemistry, biology and, finally, to what Comte was the first to term 'sociology' or 'social physics'. This idea of a hierarchy of sciences, which became so commonplace during the nineteenth and twentieth centuries, was not familiar to readers of Comte. Perhaps most controversial was the omission from the hierarchy of an autonomous science of the mind.[22] Comte himself believed that all mental phenomena could be studied either biologically (using phrenology) or sociologically.

After the completion of the *Cours*, in 1842, Comte entered what John Stuart Mill characterised as his 'second career' during which the author of the *Cours*, 'the *savant*, historian, and philosopher' was transformed into 'the High Priest of the Religion of Humanity'.[23] During the 1840s, he had formed an intense romantic, but chaste, devotion to a married woman, Madame Clotilde de Vaux. The connection was ended tragically early in 1846, when Madame de Vaux died at the age of only 32. Comte dedicated his next major work to the memory of his beloved Clotilde. The *Système de Politique Positive*, published in French between 1851 and 1854, was subtitled *Traité*

---

[20] [Thomas H. Huxley], 'Contemporary Literature: Science', *Westminster Review* (1854), **61**, 254.

[21] Auguste Comte, *System of Positive Polity, or Treatise on Sociology, Instituting the Religion of Humanity*, trans. Edward Spencer Beesly, et al., 4 vols, London, 1875–77, iv, 486.

[22] See Thomas Dixon, *From Passions to Emotions: the Creation of a secular psychological category*, Cambridge, 2003, 153–5.

[23] Mill, 'Later Speculations' (1865), 2.

de Sociologie, Instituant la Religion de L'Humanité. Comte himself attributed this new stage in his work, in characteristically self-important terms, to the inspiration of Clotilde: 'Without her I should never have been able practically to make the career of St Paul follow on that of Aristotle, by founding the universal religion on true philosophy, after I had extracted the latter from real science'.[24] Readers in Britain sympathetic with Comte's opposition to theological dogmas and superstitions, as well as with his hard-nosed scientific empiricism, found themselves confronted now with a new kind of Comtism, and one that they did not find nearly so palatable. Never one for concision, Comte developed, through four hefty volumes, a new view of a scientific society organised around both temporal and religious powers. His devotion to Clotilde found expression in the moral and religious elevation of women as spiritually superior to men, although the men retained superiority in the temporal and intellectual realms. And even in the religious realm, although the woman was to be the locus of religious authority in her own household, in society at large the priests of humanity were to be men and the High Priest of Humanity (the Pope of Positivism) was to be, naturally, Comte himself. The details of the Religion of Humanity were worked out not only in the *Système de Politique Positive* but also in the *Catéchisme Positiviste* (1852); these works appeared in English in 1875–77 and 1858 respectively.[25]

Mill, Huxley and Martineau all felt the need to disassociate themselves from the new Religion of Humanity, which was so central to Comte's scientific vision of society. In 1865 the *Westminster Review* published two long articles by Mill, which made a strong distinction between the insights of the *Cours* and the later eccentricities of the *Système de Politique Positive* and the religion of humanity.[26] Mill accused his former friend and correspondent of having, in his later work, lost touch with common sense and become colossally conceited. Despite being sympathetic to the idea of a godless religion in principle, in practice, the extent to which Comte's overweening self-confidence infused the Religion of Humanity, which, Mill wrote in 1865, 'has to be seen … to be believed', completely undermined the credibility of the work.[27] This alleged discontinuity between Comte's earlier philosophy and his later religion was vigorously denied by Comte himself and by many of his more ardent British followers. Thomas Huxley, however, also quickly decided to distance himself from

---

[24] Auguste Comte, *The Catechism of Positive Religion*, 3rd edn, trans. Richard Congreve, London, 1891, 13.

[25] See John Hedley Brooke and Geoffrey Cantor, *Reconstructing Nature: the Engagement of science and religion*, Edinburgh, 1998, 47–57; Comte (1891); Comte (1875–77); [Richard Congreve], 'Comte's Positive Philosophy', *Westminster Review* (1854), **62**, 173–94.

[26] John Stuart Mill, 'The Positive Philosophy of Auguste Comte', *Westminster Review* (1865), **27**, 339–405; Mill, 'Later Speculations' (1865). These two articles were reprinted together as a book, also in 1865: John Stuart Mill, *Auguste Comte and Positivism*, London, 1865.

[27] Mill, 'Later Speculations' (1865), 4.

Comte's later works. In his 1868 Edinburgh lecture on 'The Physical Basis of Life', and again in the pages of the *Fortnightly Review* the following year, Huxley famously dismissed the Religion of Humanity as 'Catholicism *minus* Christianity'.[28] Indeed, it was, in part, Huxley's desire to dissociate himself from this new -ism (positivism), along with older -isms such as atheism and materialism, that led him to coin the term 'agnosticism' to describe his own particular kind of secular scepticism at a meeting of the Metaphysical Society in April 1869.[29] For her part, Harriet Martineau, whose condensation of the *Cours* had been so well received, not least by Comte himself, refused to perform a similar service for either the *Système de Politique Positive* or the *Catéchisme Positiviste*, despite Comte's numerous requests.[30]

Comte's works were anomalous, not for expressing a preference for a rational religion purged of theological dogma, but for their inclusion of an immensely detailed cult of liturgy and worship around which those with a similar desire for a godless and scientific religion might organise themselves. His Religion of Humanity was very closely modelled on the hierarchical organisation and liturgy of the medieval Roman Catholic Church – from a humanistic priesthood, to a calendar of Saints, a Catechism, and a recommended regime of personal devotions. As already mentioned above, in the place of the Pope there would be the High Priest of Humanity – Comte himself. Comte had two main sources of inspiration for the creation of his religious system. On the social side, he drew on the works of the Catholic counter-revolutionary philosopher Joseph de Maistre, especially his 1819 work on the supreme sovereignty of the Pope. On the spiritual side, Comte especially admired the *Imitation of Christ*, anonymously published in 1418 and now normally attributed to Thomas à Kempis. Comte's advice to the believer in the religion of humanity was to read the *Imitation* daily, but mentally to replace the word 'God' on each occasion it was used with the word 'Humanity'.[31] By humanity, Comte meant the entire human race, past, present and future. One of Comte's favourite mottos was taken from the *Imitation*: *Amem te plus quam me, nec me nisi propter te* (May I love thee more than myself, nor love

---

[28] Thomas H. Huxley, 'On the Physical Basis of Life', in *Lay Sermons, Essays, and Reviews*, London, 1893, 121–2; *idem*, 'The Scientific Aspects of Positivism', in *Lay Sermons, Essays, and Reviews*, London, 1893, 128–50.

[29] See Bernard Lightman, *The Origins of Agnosticism: Victorian unbelief and the limits of knowledge*, Baltimore and London, 1987, 10–13, 188n.; Adrian Desmond, *Huxley: From Devil's disciple to evolution's high priest*, London, 1998, 372–5. For Lightman's more recent reappraisal of the rhetoric and propaganda surrounding the story of Huxley's coining of the term, see Bernard Lightman, 'Huxley and Scientific Agnosticism: the Strange case of a failed rhetorical strategy', *British Journal for the History of Science* (2002), **35**, 271–90.

[30] Susan Hoecker-Drysdale, 'Harriet Martineau and the Positivism of Auguste Comte', in Michael R. Hill and Susan Hoecker-Drysdale, eds, *Harriet Martineau: Theoretical and methodological perspectives*, New York and London, 2001, 186.

[31] Comte (1891), 218.

myself save for thee). The 'thee' here was, again, to be taken to be all of humanity. So, Comte and Comtean believers in Britain, the most prominent of whom were John H. Bridges, Frederic Harrison, Edward S. Beesley and Richard Congreve, went even further than those who expressed a merely intellectual preference for a scientific and humanistic religion. They put their intellectual preference into cultic religious practice and encouraged others to do likewise.[32]

The 'great problem of human life' for Comte was how to organise society so that egoism would be subordinated to altruism.[33] The aim of the Religion of Humanity was to solve this problem through social organisation and individual religious devotions. Comte also sought to tackle the problem using the tools of positive science. Two of the sciences to which Comte turned in this attempt were the biology of animal behaviour and the new cerebral science of phrenology.[34] Both of these he used to argue for the naturalness of altruism. Comte went so far as to claim that the discovery of the innateness of the altruistic sentiments ranked alongside the discovery of the motion of the earth as one of the two most important results of modern science.[35] The table in Figure 13.1 is taken from the *Positive Polity*, and illustrates Comte's phrenological arrangement of the egoistic and altruistic sentiments.

Comte presented this scientific view of human nature as a direct contradiction of Catholic doctrine (although it has been suggested that his view of Catholic doctrine was somewhat skewed by the prevalence of Jansenist rhetoric in the Catholic teachings of his youth).[36] On Comte's account, the Catholic Church taught that human nature was entirely sinful and that love and benevolence were available only through divine Grace. In other words, that humans, by nature, had no benevolence in them at all. One part of the Christian tradition with which Comte did express some sympathy, however, was St Paul's account of the internal struggle played out in the individual between the law of the flesh and the law of God (Romans 7.14–25). He saw this idea of the struggle between nature and grace as an erroneous theological version of what could, after the phrenological discoveries of Dr Franz Joseph Gall, now be recognised as a positive fact: 'The imaginary conflict between nature and grace', Comte wrote in the *Catechism of Positive Religion*, would be 'thenceforward replaced by the real opposition between the posterior mass of the brain, the seat of the personal instincts, and its anterior region, where there are distinct organs for the sympathetic impulses

---

[32] Terence R. Wright, *The Religion of Humanity: the Impact of Comtean positivism on Victorian Britain*, Cambridge, 1986.
[33] This is a recurring theme throughout the *Positive Polity*; see, for instance, Comte (1875–77), i, 73, 558–9, and ii, 172.
[34] *Ibid.*, i, 456–594.
[35] *Ibid.*, iv, 18.
[36] Jacques Maritain, *Moral Philosophy: an Historical and critical survey of the great systems*, London, 1964, 333–4.

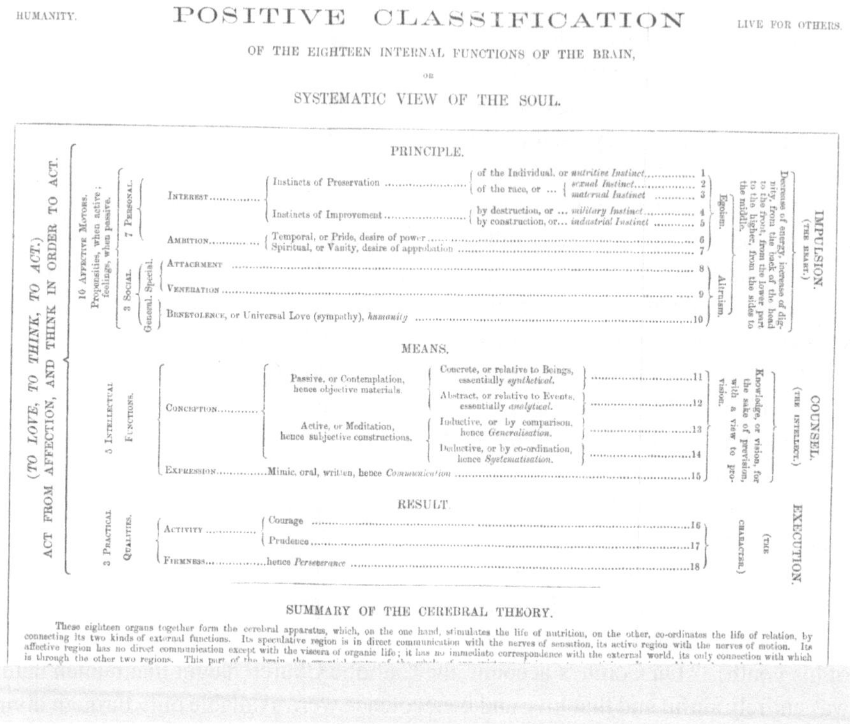

Figure 13.1 **'Positive Classification of the Eighteen Internal Functions of the Brain', taken from Auguste Comte, *System of Positive* *Polity*, *or Treatise on Sociology, Instituting the Religion of Humanity*, trans. Edward Spencer Beesly, et al., 4 vols, London, 1875–77.**

and the intellectual faculties'.[37] So, on the basis of the pervasive cooperation and self-sacrifice found in nature, and on the basis of the phrenological discoveries of Dr Gall of the location of the organs of benevolence, veneration and attachment – Comte concluded that humans were by nature inclined to have altruistic feelings. He added that both happiness and progress – the ultimate goal of positivism – would be served by the promotion of universal altruism. This alleged contrast between theological views about the corruption of human nature and the views of human nature provided by the new sciences of mind, such as phrenology, was a continued theme of Comte's

---

[37] Comte (1891), 176.

followers in Britain. John H. Bridges, writing in *The Positivist Review*, claimed that theological philosophy had always taught 'the total corruption of human nature', but that, in contrast, the 'demonstration of the fact that the impulse of unselfish love is embodied in the structure of the human brain is a discovery of modern growth, and it is one of the corner-stones of ethical science'.[38] In the Comtean view, the sciences of phrenology and sociology, in demonstrating the innateness of altruism, laid the foundations for a new society and a new scientific sort of ethics.

'Altruism' comes to England

Comte's proposition of an atheistic church and a humanistic religion both directed towards the promotion of universal altruism was by no means the only expression during the course of the nineteenth century of the desire for a religion without theology. In England, Henry Sidgwick and others, such as George Eliot, Harriet Martineau and Beatrice Webb felt just the same. They hungered and thirsted after some kind of religious orthodoxy but found they simply could not swallow it in its established Christian forms. Several key members of this group have already been very illuminatingly written about in authoritative studies by Bernard Lightman and Frank Turner.[39] Lightman has rightly emphasised the intense religiosity of these Victorian doubters as well as their intense hostility to much of Christianity (especially its more dogmatic, institutional and clerical aspects). Turner's interest, in his 1974 study, *Between Science and Religion*, was, like mine, in Victorians who found themselves alienated from traditional Christian faith but unable to convert themselves entirely to the naturalistic and atheistic world-view of scientific naturalism. An analysis of the lives and thoughts of such individuals can help to chart some of the central cultural and intellectual movements of the time. While Turner's focus was on the world-view of scientific naturalism, mine is particularly on British positivism, social science and ethics.

Many of these British doubters would have approved of Comte's own atheological articulation of the meaning of 'religion'. Religion, for Comte, was not necessarily anything to do with God or the supernatural, but was that which bound affection and intellect together within the individual, and which bound individuals together into a moral community.[40] The Cambridge divine Brooke Westcott noted that Comte was in agreement with St Augustine at least on the question of the etymological

---

[38] J.H. Bridges, 'Altruism', *Positivist Review* (1899), **7**, 177.

[39] Lightman (1987); Frank M. Turner, *Between Science and Religion: the Reaction to scientific naturalism in late Victorian England*, New Haven and London, 1974; *idem*, *Contesting Cultural Authority: Essays in Victorian intellectual life*, Cambridge, 1993.

[40] Comte (1875–77), ii, 7–11.

derivation of *religio* from the Latin *religare*, meaning to bind together.[41] In 1854, before Huxley had so decisively distanced himself from Comte and his British followers, even he could express admiration for this idea of an edifying, scientific and anti-theological religion. Writing in the *Westminster Review*, Huxley praised the eloquence of the following passage from G.H. Lewes' book on Comte's philosophy of the sciences:

> I say, therefore, that if astronomy must destroy theology, it will deepen Religion. There is no man in whom the starry heavens have not excited religious emotion ... However various the dialects and formulas into which the emotion may be translated, according to the various intellects of men, the emotion itself is constant; and the Last Man gazing upwards at the stars will, in the depths of his reverent soul, echo the Psalmist's burst – 'The Heavens declare the glory of God!'[42]

So, even the arch-sceptic Huxley, later so famous for his antipathy to Comte's humanistic religion as well as to Christianity, was not, in 1854, averse to the idea that religion could be deepened by science. Huxley might have agreed with Sidgwick's judgement, when he wrote about Comte: 'I cannot swallow his Religion of Humanity, and yet his arguments as to the necessity of Religion of some sort have great weight with me'.[43] Also in 1854, the *Clerical Journal* noted that unbelief 'clothes itself now in social respectability; it affects to have the welfare of the masses at heart, and proclaims a religion of its own'.[44] It was with this combination of concern for the welfare of the urban poor and religious unbelief, mocked by the *Clerical Journal*, that the discourse of altruism became associated. From Harriet Martineau's and Henry Atkinson's controversial 1851 *Letters on the Laws of Man's Nature and Development* to Huxley's campaigns for secular education and agnosticism later in the century, a concerted effort was under way to dissociate unbelief from decadence, immorality and revolutionary politics. The new self-sacrificing ideal of altruism fitted the bill perfectly – it was a virtue good enough for any respectable unbeliever, and one which was, at least to some extent, independent of the language and metaphysics of Christianity. It was claimed by some that this sort of unbelieving altruism was, in fact, morally superior to Christianity, which was an essentially selfish system based on each individual's desire for infinite reward and fear of eternal punishment.[45]

---

41 Brooke F. Westcott, 'Aspects of Positivism in Relation to Christianity', *Contemporary Review* (1868), **8**, 374.

42 [Huxley] (1854), 257; [Lewes] (1852), 92.

43 Schneewind (1977), 43.

44 Quoted in Edward Royle, *Victorian Infidels: the Origins of the British secularist movement, 1791–1866*, Manchester, 1974, 157.

45 For example, H.G. Atkinson and Harriet Martineau, *Letters on the Laws of Man's Nature and Development*, London, 1851, 246–8; Comte (1891), 22, 216.

The term altruism thus became a valuable tool in the project of creating a cluster of new social identities from the 1850s onwards – the respectable unbeliever, the secular philanthropist, the honest doubter, the scientific agnostic. The rise of 'altruism' from the 1850s to the 1890s coincided with the coining of the term 'agnosticism' (1869), the election of the first avowedly atheist Member of Parliament (Charles Bradlaugh, in 1880), and the election of the first unbelieving president of the Royal Society (Thomas Huxley, in 1883). Beatrice Potter (later Beatrice Webb) was another respectable unbeliever who was attracted to the language of altruism. In 1884, at the age of 26, she opened a new notebook for herself, and copied out the following passage from Comte: 'Our harmony as moral beings is impossible on any other foundation but altruism. Nay more, altruism alone can enable us to live in the highest and truest sense. To live for others is the only means of developing the whole existence of man'.[46] Although Potter resisted the attempts of Frederic Harrison to persuade her to become a member of the Positivist Church of Humanity, her adoption of the language of altruism was another indication of how Comte's combination of social science with atheistic religiosity could appeal to a powerful set of nineteenth-century aspirations.

But how was this new ethical ideal of altruism received in the world of Victorian Christianity? Some felt that this was, if you like, merely Christian ethics in disguise – that altruism and *agape* were identical.[47] There were also those who felt that altruism, because of its association with an atheistic world-view, was nothing more than a mockery of true Christian love. Others combined elements of both these views.

Brooke Westcott, who was unusual amongst Anglican churchmen in taking a serious and sympathetic interest in Comte's work, saw a deep affinity between positivist and Christian teachings. In an 1868 article for the *Contemporary Review* on Comte's religious ideas, Westcott wrote: 'Now these grand and far-reaching ideas of the continuity, the solidarity, the totality of life, which answer equally to the laws of our being and the deepest aspiration of our souls, are not only reconcilable with Christianity, but they are essentially Christian'.[48] Henry Drummond's attempt to Christianise the Comtean picture of nature and altruism (the capitalisation was his own), in his best-selling 1894 *Ascent of Man*, is particularly striking. He sought to effect a fusion of the languages of positivist science and Christian theology. He described evolution as a love-story in which altruism – self-sacrificing love – was present from the earliest stages. Both selfish egoism and altruistic love are great in nature, he wrote, but – echoing St Paul – he concluded that 'the greatest of these is Love'.[49]

---

[46] Quoted in Beatrice Webb, *My Apprenticeship*, London, 1926, 149; this is her own translation from the *Cathéchisme Positiviste*; for Congreve's translation, see Comte (1891), 215.

[47] A recent book by Colin Grant also argues along these lines; Grant (2001).

[48] Westcott (1868), 383.

[49] Henry Drummond, *The Lowell Lectures on the Ascent of Man*, London, 1899, 341.

There were other Christian writers who, unlike Westcott and Drummond, were less keen to see this new term 'altruism' identified with Christian love. In the Bampton Lectures of 1877, Charles Row mocked the new Comtean movement. 'These modern times', he told his Oxford audience, 'have set up a phantom called the religion of humanity, whose great moral principle is altruism, or the sacrifice of the self to the idea of human nature ... – a mere Caricature of Christianity. But it is powerless! Where is its army of self-sacrificers? It stamps on the ground, but no legions appear at its bidding'.[50] In fact, many other Christian responses combined elements of both of these two positions. The resulting view was that altruism was indeed the ethical ideal taught by Jesus Christ, but that the way to achieve universal brotherly love was through God and salvation, rather than through humanity or social organisation. Comte's sociocracy could not deliver the goods, these Christian writers thought, any more than the eugenic programme of breeding altruists, which some were proposing could.[51] F.W. Farrar articulated this view in his Hulsean Lectures at Cambridge in 1870. 'Live for others', is indeed a grand motto, Farrar said, but he did not see the need for Comte's new terminology. 'Is "altruism" a sweeter or better word than charity?' he asked; 'or has the bare and naked formula "*Vivre pour autrui*" a charm which has been lost from the old commandment, "Thou shalt love they neighbour as thyself"?' Furthermore, by detaching love of neighbour from love of God, Farrar said, the religion of humanity took away the very basis and reality that made such love possible.[52] As William Boyd Carpenter put it in 1887, four years after delivering his sermon at Windsor about John Brown: 'The principle of Altruism will, no doubt, win its way in the world ... But when that vision is realised, it will not be through the unaided sentiments of man – the holy city of universal brotherhood is a city whose builder and maker is God'.[53]

Beyond 'Science' and 'Religion': The Politics of Altruism

In this essay I have been making use of the categories 'science' and 'religion' in my analysis of the significance of the adoption of the new Comtean terminology of

---

[50] C.A. Row, *Christian Evidences Viewed in Relation to Modern Thought*, London, 1877, 106.

[51] James Cotter Morison (author of *The Service of Man*, London, 1887) was identified by one writer as an advocate of such a breeding programme: Moritz Kaufmann, *Egoism, Altruism, and Christian Eudaimonism*, London, 1890, 25.

[52] F.W. Farrar, *The Witness of Christ to History, Being the Hulsean Lectures for the Year 1870*, London and New York, 1871, 144–6.

[53] W. Boyd Carpenter, *The Permanent Elements of Religion: the Bampton lectures of 1887*, London, 1889, 219–21.

'altruism' in various Victorian contexts. The application of these categories is often problematic and potentially misleading, as many historians have noted.[54] Even though they are, in this case, certainly categories used by the historical actors under discussion, it is still useful to think about ways to look beyond this sort of analysis, in order to avoid projecting contemporary ideas about the meaning of these terms back into contexts in which they do not belong. As I have already suggested, there are many aspects of Victorian intellectual life that resist simple categorisation as either 'science' or 'religion'. Comte's *Positive Polity* was at the same time both religious and scientific in its methods and conclusions; as its sub-title revealed, it aspired to be the founding document both of a new science (sociology) and of a new religion (the Religion of Humanity). A second way that our contemporary assumptions about science and religion might be misleading is if we were inclined to think that religion could be a sound basis for an ethical system but that science could not. Ultimately, when it came to ethics, science and religion were in a similar position. In both cases ethical prescriptions were arrived at by a process of reading – either the book of nature or books of scripture. Reading nature and reading scripture were, and still are, both highly interpretative activities, attempted reconstructions of a posited unseen that lies behind a veil (of phenomena or of words). Nature and scripture could each be appealed to as sources of moral authority, but which was chosen may not ultimately have been the most important question; neither could be entirely compelling. Looking at Victorian discussions of altruism reinforces this realisation of the plasticity of both nature and scripture; this should perhaps lead to a certain amount of suspicion about any claims we come across in our own time to derive moral prescriptions directly from either scientific or religious authority. It might also encourage us to wonder how contemporary claims about relationships between nature and ethics or scripture and ethics will seem to future historians.

Looking beyond the scientific and religious explanations and justifications of arguments about human altruism, then, it seems that at root what was at stake were primarily neither scientific nor religious issues but fundamentally political questions about what constituted the ideal society, about who was a member of one's moral community, and about how that community should be organised. This claim – that debates about the meaning of 'altruism' were fundamentally political – is also supported by the fact that by the last two decades of the nineteenth century the term 'altruism' had taken on another distinctive new meaning. It had become widely associated with socialism, and with cooperative communes in Britain and America. In the rather

[54] See, for instance, James R. Moore, 'Speaking of "Science and Religion" – Then and Now', *History of Science* (1992), **30**, 311–23; Turner (1993); David B. Wilson, 'On the Importance of Eliminating *Science* and *Religion* from the History of Science and Religion: the Cases of Oliver Lodge, J.H. Jeans and A.S. Eddington', in Jitse M. van der Meer, ed., *Facets of Faith and Science, Vol. 1: Historiography and modes of interaction*, Lanham, MD, 1996, 27–47.

conservative London *Spectator*, anxieties were being expressed about the ideological extravagance and fanaticism of the new socialist advocates of altruism.[55] According to one contributor, writing with alarm in 1897, it was being claimed 'in a thousand pulpits that Christ taught altruism, and altruism as understood by those who have accepted the semi-Socialist or Socialist theory now so prevalent'.[56] Also in 1897 a comic novel, *An Altruist*, appeared, which spoofed this new generation of idealistic altruists. The central character, one Wilfrid Bertram, is described as 'an altruist, a collectivist, a Fourrierist, an Engelist, a Tolstoi-ist'. One listener to Bertram's creed of collectivity, when asked whether the doctrine has been made clear, responds: 'Nothing could be clearer than what you've said. … Nobody is to have anything they can call their own, and everybody who likes is to eat from one's plate and bathe in one's bath'.[57] One such commune, actually established in California in the 1890s, was even named 'Altruria'. This colony was meant to be a refuge for those tired by competitive, selfish modern society. The entry requirements were 50 dollars and a moral character; the community met on Sundays for a humanistic religious service involving music and discussion; their ideology was one of brotherhood and mutuality. There was complete freedom of religious opinion; however, it was a missionary foundation, committed to founding new settlements on the same model. Like many Utopian schemes, however, this one failed financially.[58]

By the 1890s, then, the association of altruism with Comtean sociology was somewhat looser. The religious connotations of the term, however, continued to be both humanistic and Christian even in this new era of socialist altruism. Thus the journalist and socialist campaigner Robert Blatchford could write in a pamphlet in 1898: 'Altruism, which is the embodiment of the command "Love they neighbour as thyself," seems to have originated in the teachings of Christ, but has only attained important development in comparatively modern times'.[59] Blatchford went on to describe the spread of altruism as 'the most important consummation in the progress of social evolution'.[60] The faith he hoped other socialists shared with him, in the importance of spreading universal brotherly love, Blatchford said, was not founded on economic science but must have as its basis 'a religion of passionate love for humanity, a religion of service and sacrifice – a religion of Altruism'.[61]

---

[55] Anon., 'Prodigality and Altruism', *Spectator* (1884), **57**, 375–6; Anon., 'The Extravagance of Altruism', *Spectator* (1892), **68**, 671–2; Anon., 'The Weak Point in Altruism', *Spectator* (1897), **79**, 515–16.
[56] Anon. (1897), 515.
[57] Ouida [pseudonym, Louise de la Ramée], *An Altruist*, 2nd edn, London, 1897, 15.
[58] Edward B. Payne, 'Altruria', *American Magazine of Civics* (1895), **6**, 168–71; Morrison I. Swift, 'Altruria in California', *Overland Monthly* (1897), **29**, 643–5.
[59] Robert Blatchford, *Altruism: Christ's Glorious Gospel of Love against Man's Dismal Science of Greed (Clarion Pamphlet No. 22)*, London, 1898, 3.
[60] *Ibid.*, 6.
[61] *Ibid.*, 8.

The socialist altruists, like Comte and others before them, recognised that a scientific explanation of the springs of selfless behaviour could not in itself supply a practical solution to Comte's 'great problem of human life' – how to subordinate egoism to altruism. What was needed was a political programme. While Comte's suggested solution had been communal humanistic worship and social management, George Eliot's, for example, had been improved secular education, the eugenicist James Cotter Morrison had proposed a programme of selective breeding of altruists, and the new socialist altruists of the 1880s and 1890s favoured simple communal living centred around socialist ideals as the practical way to live out this idea.[62] The humanistic ideal of Comte's envisaged anti-democratic and hierarchical religious society had become the name for a new sort of democratic and idealistic socialism. 'Altruism' had thus entered a new stage in its own development, beyond the theological, the metaphysical and, even, the positive. The very fact that the ideal of 'altruism' could be used both as a device to promote a conservative Christianity which kept the servants in their place by honouring their self-sacrifice (as Boyd-Carpenter had done in his sermon at Windsor Chapel), and also as a rallying call for respectable unbelievers and, even, for radicals and communitarians, illustrates that, as ever, scientific theories (such as Comte's phrenological and sociological accounts of altruism and egoism) could be appropriated within an almost indefinitely wide range of political projects.

To return, finally, to more recent sociobiological treatments of altruism, is there a similar political dimension to the debate? I believe that there is. Richard Dawkins prefaced his famous pronouncement in *The Selfish Gene* about the need to teach altruism with the following words:

> My own feeling is that a human society based simply on the gene's law of universal ruthlessness would be a very nasty society in which to live. ... Be warned that if you wish, as I do, to build a society in which individuals cooperate generously and unselfishly towards a common good, you can expect little help from biological nature. Let us try to *teach* generosity and altruism, because we are born selfish ...[63]

Arguments about altruism, whether presented as scientific, religious, or both, in the Victorian period or more recently, have always involved fundamentally political proposals about the ideal human society, about the boundaries of moral communities, and about the mechanisms of social improvement.

---

[62] [Eliot] (1856), 53–6. On Cotter Morison, see n. 52 above.
[63] Dawkins (1989), 3.

Chapter 14

The Radiometer and its Lessons: William Carpenter *Versus* William Crookes

William H. Brock

> At Gabriel College there was a very holy object kept on the high altar of the Oratory, covered with a black velvet cloth. … At the height of the invocation the Intercessor lifted the cloth to reveal in the dimness a glass dome inside which there was something too distant to see, until he pulled a string attached to a shutter above, letting a ray of sunlight through to strike the dome exactly. Then it became clear: a little thing like a weathervane, with four sails black on one side and white on the other, began to whirl around as the light struck it. It illustrated a moral lesson, the Intercessor explained, for the black of ignorance fled from the light, whereas the wisdom of white rushed to embrace it.[1]

In that parody of a Christian ritual, Philip Pullman transforms the magical effect of a scientific instrument, the radiometer, into a holy relic and satirises religious homilies by drawing a moral message from its mysterious action in the presence of light.

Unknown to Pullman, probably, when the radiometer was first constructed in the early 1870s it must have seemed an extraordinary and mysterious object and one that might, indeed, link the worlds of material scientific reality with the mysterious and religious unknown. It also brought about a mighty row of the kind that Victorian men of science delighted in.

David Knight has noted how great rows were characteristic features of nineteenth-century science.[2] Soon after he and I completed a joint paper on the atomic debates in 1964, I received an invitation to write the entry on William Crookes for the *Dictionary of Scientific Biography*.[3] Crookes had many spectacular rows during his

---

[1] Philip Pullman, *Northern Lights*, Scholastic Ltd, London, 1995; pbk edn 2001, 149. Cited by permission of the author and Scholastic Children's Books.

[2] David Knight, *The Age of Science. The Scientific world-view in the nineteenth century*, Oxford, 1986, 7.

[3] W.H. Brock and D.M. Knight, 'The Atomic Debates: "Memorable and interesting evenings in the life of the Chemical Society"', *Isis* (1965), **56**, 5–25; W.H. Brock, 'William Crookes', *Dictionary of Scientific Biography*, iii (1971), 474–82.

long career. None, perhaps, was more spectacular than the six-year running feud with the physiologist William Benjamin Carpenter that forms the subject of this essay.

Born into a famous Unitarian family, Carpenter had enjoyed a sound medical education in Bristol, London and Edinburgh. As a Unitarian, Carpenter was a committed deist who believed God controlled nature by natural laws. Carpenter had been born in 1813, and was nearly 20 years older than Crookes. He had been a successful medical practitioner in Bristol before moving to London, and had written his first book at the age of 26, the highly original *Principles of General and Comparative Physiology*.[4] Crookes on the other hand never wrote an original book apart from his popular *Diamonds*.[5] Moving to London in 1844, Carpenter had quickly found positions at University College, the London Hospital, and the Royal Institution and forged a financially satisfying career as lecturer, researcher on comparative physiology, microscopy and marine biology, essayist, reviewer, editor and critic. From 1856 onwards, he gained complete financial security in the administrative post of registrar to the University of London. He also shared Crookes' ability as an editor, having edited the *British & Foreign Medico-Chirurgical Review* for five years from 1847 to 1852, and even a critical *Spiritual Messenger, a Magazine devoted to Spiritualism, Mesmerism and other Branches of Psychological Science* in 1858–59.

When not engaged in psychological and physiological research, organ playing at his Unitarian church, or administering the University of London, Carpenter spent his time hunting out fraud and learning from the abnormal. In an important essay in the *Quarterly Review* in 1852 he made a scintillating attack on quack mesmerists, while supporting the theory of suggestion, or 'hypnosis', that had been propounded by the Mancunian doctor, James Braid. According to Carpenter, the human mind could easily become obsessed by a 'dominant idea' that was placed there either by the hypnotist's suggestions or, more commonly, through autosuggestion. He further argued, foreshadowing Freud, that such dominant ideas could produce 'unconscious cerebrations' or notions of expectancy. In this way the deluded person was unconsciously led into abnormal behaviour or into ridiculous beliefs. Table-turning was, as Faraday had experimentally demonstrated, produced by muscular expectancy induced by the dominant idea or conviction that something dramatic was going to happen and that spirits were communicating. Where trickery and tomfoolery could be ruled out, the startling events of the mid-Victorian séance were to be explained by expectancy: the participants at the séance unconsciously made the things happen themselves. Thus as early as 1852 Carpenter had gone out of his way

---

[4] W.B. Carpenter, *Principles of General and Comparative Physiology*, London, 1839. On Carpenter, see *DSB* entry by K. Bryn Thomas.

[5] Sir William Crookes, *Diamonds*, London, 1909; this was based on a British Association lecture given at Kimberley in South Africa, June 1906. Crookes' many other book publications were all translations or editions of other authors' work.

to attack the incipient spiritualist movement. As Alison Winter has shown, Carpenter was enormously successful in propagating the view that much of human behaviour was involuntary.[6]

Crookes never enjoyed such early fortune. Degreeless, he ran through a succession of short-term jobs in the 1850s, failed to find a university of administrative post, and was forced to earn his living as editor and consultant, neither of which gave him a lot of time for research.[7] Yet, like Carpenter, Crookes was elected to the Royal Society at the age of 31. But whereas Carpenter's election in 1844 was based on a solid contribution to physiology, Carpenter felt that Crookes' election in 1863 for the discovery of thallium had been largely undeserved and was a matter of controversy.

Although passing comments on spiritualism in *Chemical News* in the 1860s show Crookes was initially scornful of the movement, he appears to have changed his mind in 1869 following the death at sea of a beloved younger brother, Philip Crookes. The death seems to have affected the whole of Crookes' very large family so intently that they all sought solace in attending séances. Most intriguingly, Alfred Crookes, his 47-year-old bookseller stepbrother by his father's first marriage, took control – that is, he became the manager – of the young medium Frank Herne. At one such séance Herne spoke so effectively to William Crookes about the deceased Philip that Crookes was moved to tears. Convinced of the validity of his experience, he decided to use his scientific expertise to investigate spiritualist phenomena.[8]

The investigation of curious anomalies thrown up during his exacting determination of the atomic weight of thallium between about 1862 and 1870 led Crookes and his apprentice Charles Gimingham to the construction of the eye-catching and puzzling light-mill or radiometer. He first demonstrated this to the scientific public, namely the Royal Society, in April 1875. But, as early as 1870, he had used a prototype privately at séances in the hope that it might help elucidate what he had already christened the 'psychic force' possessed by some mediums. Indeed, here might be an instrument that separated genuine mediums from prestidigitators and illusionists. We know from a letter he addressed to the astronomer William Huggins and from reports made by

---

[6] Alison Winter, *Mesmerized. Powers of mind in Victorian Britain*, Chicago and London, 1998.

[7] W.H. Brock, 'A British Career in Chemistry: Sir William Crookes (1832–1919)', in David Knight and Helge Kragh, eds, *The Making of the Chemist. The social history of chemistry in Europe 1789–1914*, Cambridge and New York, 1998, 121–9.

[8] W.T. Stead, 'Who May this Be?"', *Borderland. A Quarterly Review and Index* (1896), **3**, 191. Herne must have been only 18 or 19 when managed by Alfred Crookes. He later became an early associate of Florence Cook and was 'exposed' as a fraud in 1875. See Janet Oppenheim, *The Other World. Spiritualism and psychical research in England, 1850–1914*, Cambridge, 1985. Another brother, Walter, also attended séances, as well as Crookes' children, Henry and Alice, his wife and mother-in-law, Mrs Humphrey.

Francis Galton to Charles Darwin that in 1871 Crookes believed that the radiometer effect had provided him with a sensitive indicator of the presence of a strange force that mediums used.[9]

In July 1870 Crookes published an article entitled 'Spiritualism Viewed in the Light of Modern Science' in his own journal, the *Quarterly Journal of Science*.[10] This was a serious, almost sceptical, review of psychic phenomena, but reached the conclusion that, trickery aside, there was a residuum of phenomena 'inexplicable by any known natural laws'. It was these non-conjuring or non-trick phenomena that needed testing under laboratory conditions. It would appear that the scientific community and critics of spiritualism initially welcomed Crookes' promise to make a serious investigation. The dominant belief of the scientific community was that Crookes would condemn spiritualism as fraudulent or, at least, declare it mistaken. The believers in spirit communication with the dead (though not all) trusted that Crookes would validate their experiences However, when a year later he offered a report to the Royal Society, it was summarily rejected; whereupon Crookes again published it in his own journal.[11] In this 'Experimental Investigation of a New Force', he claimed to have found a subject in the Scottish-American medium, David Dunglas Home, who could be subjected to laboratory conditions and yet still produce strange phenomena.

Crookes described two basic experiments. In the first, Home was able to play a melody on an accordion whilst his hands were prevented from touching the keys. In a second, more quantified, experiment, one end of a plank of wood was suspended from a spring balance whilst its other end just rested on a table. Home's fingers were placed lightly on the table-end of the plank before the fulcrum and, on exerting his 'psychic force' an oscillating depression of the spring balance was recorded by the onlookers. Crookes attributed this not to any surreptitious movements on Home's part, but to a genuine flow of nervous energy, or 'psychic force', from Home's body. He noted with approval that Home seemed exhausted after the experiment, as if the principle of energy conservation had not been broken.

Crookes had invited both secretaries of the Royal Society, William Sharpey and George Stokes, to attend the experiments with Home, but Sharpey had declined and Stokes stated that while he would gladly inspect the apparatus he too would not wish to attend personally. Three months later, in October 1871, Crookes published 'Some

---

[9] Crookes to Huggins, 6 November 1871, cited in E.E. Fournier d'Albe, *The Life of Sir William Crookes*, London, 1923, 227; Galton to Darwin, 28 March 1872, cited by R.G. Medhurst and K.M. Goldney, 'William Crookes and the Physical Phenomena of Mediumship', *Proceedings of the Society for Psychical Research* (1964), **54**, 25–157, 41–4.

[10] *Quarterly Journal of Science* (July 1870), **7**, 316–21; reprinted in W. Crookes, *Researches in the Phenomena of Spiritualism*, London, 1874, with many reprintings.

[11] W. Crookes, 'Experimental Investigation of a New Force', *Quarterly Journal of Science* (July 1871), **8**, 339–49, also reprinted in Crookes (1874).

Further Experiments on Psychic Force', prefacing it pugnaciously with a stirring quotation by Galvani:

> I am attacked by two very opposite sects – the scientists and the know-nothings. Both laugh at me – calling me 'the frogs' dancing master'. Yet I know I have discovered one of the greatest forces in nature.[12]

These three *Quarterly* essays, together with a number of more startling reports of his investigations of the physical manifestation medium Florence Cook that appeared in the weekly *Spiritualist* in 1874, formed the basis for Crookes' reputation as either a courageous scientist investigating the unknown, or a garrulous fool who had been hoodwinked and deluded by his own megalomania.[13]

Carpenter was not Crookes' only critic. John Parsons Earwaker, a lawyer and Fellow of Merton College, Oxford, related Crookes' work on psychic powers to the history of mass delusions. Crookes', as well as the astronomer William Huggins', uncritical investigations threatened to turn the honorary letters FRS into 'Follower of Rampageous Spiritualism'. Earwaker 'blushed' to think that Crookes and Huggins could be taken in by two poorly designed experiments. Why go to great lengths to build a Faraday cage for the accordion experiment with Home and then place it *under* the table rather than in the middle of the room? Earwaker extracted a good deal of fun from the poor design and reporting of this experiment. As for the plank experiment in which Home apparently caused a depression of a spring balance, Earwaker noted that the experimental setup turned on Home not moving the table itself, 'the one precaution not attended to'.[14] This would, indeed, be the most obvious interpretation and one closely related to familiar table moving in séances. (However, further experiments that Crookes only revealed to the Society for Psychical Research in 1889 appear to have guarded against the possibility that Home manipulated the table.[15])

Earwaker's most telling point was that irrelevant experimental data were collected (such as the temperature of the room) that gave the tests a pseudo-scientific dressing, while more obvious precautions were totally ignored because Crookes and Huggins presumed Home an honest gentleman. As for Crookes' complaint that other scientists, apart from Huggins, Cromwell Fleetwood Varley and Alfred Russel Wallace, had refused to investigate or support his investigations, Earwaker declared this was

---

[12] W. Crookes, 'Some Further Experiments on Psychic Force', *Quarterly Journal of Science* (October 1871), **8**, 471-93, reprinted in Crookes (1874).

[13] The most convenient source for Crookes' reports on Florence Cook is Crookes (1874).

[14] J.P. Earwaker, 'Letter on Psychic Force', *Nature* (3 August 1871), **4**, 278–9; 'Mr Crookes' New Psychic Force', *Popular Science Review* (1871), **10**, 356–65, 360.

[15] W. Crookes, 'Notes on Séances with D.D. Home', *Proceedings of Society for Psychical Research* (1889), **6**, 98–127.

untrue. Crookes and others had been encouraged to investigate psychic powers at the Edinburgh meeting of the British Association for the Advancement of Science in 1871 when George Stokes had even proposed the formation of a British Association Committee. Nothing had come of this because it had been discovered that another independent committee had been formed in St Petersburg. This committee had seated Home at a glass table on which was placed a lamp with a reflector. It transpired that Home had been unable to increase the weight of the plank suspended by the spring balance. On this occasion Home had suspiciously excused himself on the grounds of fatigue.[16]

In an article by the American physician Peter H. Vander Weyde (1813–95), who claimed to have spent 20 years investigating 'so-called spiritualism', the *Scientific American* dismissed Home's phenomena as those of a first-class prestidigitator, who took advantage of the contemporary fashion for belief in spiritualism. The accordion playing was a well-known ventriloquist's trick that involved placing a miniature harmonica in the mouth; the Faraday wire cage served only to hide additional means of support for the accordion; the spring balance experiment was done by an electromagnet, etc. Any illusionist who had been present would have been able to spot other ways in which the tricks were done. Crookes' gullibility was like that of the chemists who invented phlogiston as an explanation for combustion or of the physiologists who evoked a vital force to explain the mystery of life. Crookes should have used a chemical balance (Weyde did not explain how he could) rather than a spring balance since the latter was always open to trickery, as all grocers knew. To add insult to injury, the plank experiment was not original, but adopted from the notorious experiments that the Philadelphia chemist Robert Hare had conducted with a child medium in 1853. It was widely known in America, Weyde asserted, that the 72-year-old Hare had gone mad.[17]

Another American critic was the mining engineer Henry Wurtz, who had crossed Crookes' path ten years before when both men were developing the silver amalgamation process for the recovery of silver in mining operations. Writing in the *American Gas-light Journal*, Wurtz explicitly condemned the term 'psychic force' as implying trickery, and whimsically attacked the idea of revealing how such things were done as spoiling the wonder of the child's belief in a magician's tricks. Both Vander Weyde's and Wurtz's articles were reprinted for British readers in the popular weekly *English Mechanic* during the autumn of 1871. In an editorial the journal stated the obvious, that either Crookes had been deceived or the phenomena were real. In

---

[16] For the context of Home's Russian experience, see Richard E. Rice, 'Mendeleev's Public Opposition to Spiritualism', *Ambix* (1998), **45**, 85–95.

[17] Robert Hare, *Experimental Investigations: the Spirit manifestations*, New York, 1855; Frank Podmore, *Modern Spiritualism*, 2 vols, London, 1902, i, 233–5.

either case, it was a pity that a proper committee of inquiry had not been set up by the Royal Society.[18]

Further controversy was created by an anonymous letter (in fact by the astronomer William Noble) to the *English Mechanic* suggesting that John Spiller, who had assisted at one of the accordion-playing séances, had found out how Home had done it. Noble demanded that Spiller spill the beans.[19] Crookes replied pointing out that Spiller was not his assistant (that was Charles Gimingham) and that Spiller had not been present at the experiments the previous July, though he agreed that Spiller and Home had met socially at both Crookes' house in Mornington Road and at that of the eminent lawyer Serjeant Cox. Crookes angrily demanded to know the source of these rumours.[20]

Noble replied that he had heard about Spiller's accusation independently from a Cambridge graduate and a secretary of one of the learned societies (probably the Royal Astronomical Society). Both men had also said uncomplimentary things about Crookes' scientific attainments.[21] In reply to this Crookes sent Noble (via the editor) a copy of his letter to *The Echo* newspaper of 10 November which revealed that Spiller truly had accused Home of jigglery and fraud, and Crookes of gullibility. At this point in the debate Spiller decided he had better enter the *English Mechanic* debate. On the 1 December 1871 he admitted that he had attended séances on 19 April (at Crookes') and 25 April (at Cox's). Crookes had not welcomed his presence at the latter séance at which he had refused to put his name to a document describing the phenomena witnessed, at which juncture either Crookes or Home (it is unclear which) had threatened legal proceedings on the grounds that Spiller was libelling the personal character of Crookes, Huggins and Cox. Spiller had been particularly struck by the huge size of a platinum locket that dangled from Home's watch chain and suspected it played a role in the proceedings.[22]

Now that this information was public knowledge, Noble seized on the threat of legal proceedings as making any scientific investigation of psychic phenomena impossible. In reply, Crookes angrily attacked his old friend Spiller (they had been

---

[18] 'Psychic Force' (editorial), *English Mechanic* (13 October 1871), **14**, 85; P.H. Vander Weyde, 'On the Psychic Force', *English Mechanic* (13 October 1871), **14**, 87–8; Henry Wurtz, 'American Opinions of Psychic Force', *English Mechanic* (20 October 1871), **14**, 110–11.

[19] A fellow of the Royal Astronomical Society [W.H. Noble], 'Psychic Force', *English Mechanic* (3 November 1871), **14**, 168. Captain William Noble (1828–?), FRAS, was a regular contributor to the *English Mechanic*.

[20] W. Crookes, 'Psychic Force', *English Mechanic* (10 November 1871), **14**, 200. Gimingham had become Crookes' apprentice in May 1869. Edward William Cox (1809–79), barrister and periodicals editor, was a keen investigator of psychic phenomena. See *Oxford DNB*.

[21] *English Mechanic* (17 November 1871), **14**, 219–20.

[22] W, Crookes, 'Psychic Force', *English Mechanic* (24 November 1871), **14**, 253; W. Noble, *English Mechanic* (1 and 8 December 1871), **14**, 273, 299; J. Spiller, *English Mechanic* (1 December 1871), **14**, 273. See also *The Echo* (31 October, 7–8 November 1871).

students together at the Royal College of Chemistry) as propagating a tissue of lies and falsehoods.[23] The legal threat was a total perversion of the truth; the reflecting power of Home's locket was a red herring; Crookes had physically examined the locket with a photometer and found nothing suspicious. He would be publishing further details in a pamphlet to be issued by Longman. This was his *Psychic Force and Modern Spiritualism: a Reply to the Quarterly Review and other critics* published in January 1872. In order to maximise the possibility that it would be read by the scientific community, Crookes had copies bound up with the advertisement sheets of the January 1872 issue of his *Quarterly Journal of Science*.

Meanwhile, just before Christmas 1871, Spiller had replied that Crookes was fudging the real issue, namely the reality or spuriousness of psychic force. Spiller accepted Crookes' explanation that he and Huggins had merely attended dress rehearsals and not full-dress performances of the experiments with Home, and that he should have made allowances for this. Yet the fact remained that Crookes had sent him a statement to sign that purported to describe the events witnessed by all of them on 25 April. Spiller quoted what Crookes had asked him to sign and drew attention to the two points he had been unable to accept. He pointed out that he had denied the authenticity of the report in a letter to Crookes dated 3 May and so could not be accused of only now (in November/December 1871) rejecting it. What, he wondered, did Huggins, the other scientific participant, think? Spiller closed his letter by accepting Crookes' word that he had not intended to give the impression that he had threatened to take legal action.[24]

At this juncture, Passmore Edwards declined to publish any more letters from either Crookes or Spiller, stating that the dispute had become a personal slanging match that had nothing to do with the existence or non-existence of psychic force. The disputants had lost sight of the really interesting possibility that this newly posited force had been detected in a living person, as opposed to a spiritualist phenomena caused by dead spirits. This had been one of Spiller's original points – that his open mind on the existence of a psychic force had been overturned by the tomfoolery of table thumping during the experiments.

Despite Edwards' injunction, he allowed other correspondents to continue the debate. Noble suggested that Home should make himself available to a party of sceptical scientists, suggesting the names of Tyndall, Carpenter, Huxley, B.W. Richardson and a magician named Hermann. Needless to say, this did not come about. On the other hand, the electrician Cromwell Varley came to Home's defence with a

---

[23] Noble, *English Mechanic* (8 December 1871), **14**, 299; W. Crookes, *English Mechanic* (15 December 1871), **14**, 327.

[24] J. Spiller, 'Psychic Force', *English Mechanic* (22 December 1871), **14**, 357–8. Edwards closed the debate at the end of Spiller's letter.

letter to the *English Mechanic* in January 1872, saying that Home had always expressed willingness to be investigated. Unfortunately, the conditions imposed by sceptics like Faraday and Tyndall had been impossible for him to accept.[25] Varley himself had been initially sceptical, but had come round to accept the phenomena as genuine.

Edwards was not impressed by this intervention and in a long editorial note appended to Varley's letter he said that men of science could never be satisfied while the phenomena were only tested in private houses in darkened, heavily curtained rooms. Curiously, despite the editor's earlier embargo, Spiller had the last word on the matter in the *English Mechanic* in March 1872. He publicised the fact that in *The Athenaeum* of 6 January Huggins had withdrawn his support for Crookes' account; moreover, he revealed, as reported by the *Daily Telegraph,* the Council of the Royal Society had rejected Crookes' revised paper on the Home experiments on 18 January. The Chemical Society had also refused a presentation copy of Crookes' pamphlet *Psychic Force and Modern Spiritualism*. Finally, for good measure, he rubbed in the fact that the paper on Home that Crookes had offered to the British Association meeting in Edinburgh in August 1871 had also been rejected.[26] He noted how unanimous the press had been against Crookes. *The Times, Daily News, Spectator, Quarterly Review, Popular Science Review* and the *Chemist & Druggist* had all reported unfavourably on Crookes' investigation. Crookes' case for the existence of a psychic force, he concluded, was unproven.

Chemist & Druggist, the pharmacy journal, in reviewing Carpenter's essay on spiritualism,[27] had wondered why attendances at spiritualist séances were confined to no more than 20 or so people when the implications, if true, were so weighty. *Chemist & Druggist* was convinced it was all humbug. While catching the Davenport Brothers in knavery was not a reason for believing that Home was in the same line of deceit, it was a warning to be sceptical.[28] Crookes had been right to investigate Home's claims, but the public was also right to remain suspicious. It saw the debate between Carpenter and Crookes as a literary tournament, and agreed with Carpenter that whatever wriggling Crookes did, he had become converted to Spiritualism, and

---

[25] C.F. Varley, 'Psychic Force', *English Mechanic* (19 January 1872), **14**, 454–5. See Richard Noakes, 'Telegraphy is an Occult Art: Cromwell Fleetwood Varley and the diffusion of electricity to the other world', *BJHS* (1999), **32**, 421–59. Note also the reprint from *Journal Franklin Institute* of further criticisms by Vander Weyde, *English Mechanic* (12 January 1872), **14**, 425–6.

[26] J. Spiller, 'Psychic Force', *English Mechanic* (1 March 1872), **14**, 615.

[27] [W.B. Carpenter] 'Spiritualism and its Recent Converts', *Quarterly Review* (October 1871), **131**, 301–53.

[28] 'The Controversy on Psychic Force', *Chemist & Druggist* (15 February 1872), **13**, 47–8. The American brothers Ira and William Davenport were illusionists who apparently produced spiritualistic phenomena while tied to chairs inside a locked wooden cabinet. The rival stage magicians John Maskelyne and George Cooke revealed the secret of their trickery during their European tours of 1864 and 1868. See Podmore (1902), ii, 55–61.

therefore biased. It believed Carpenter's explanation of spiritualists' claims of contact with the dead in terms of unconscious cerebration and unconscious muscular action (or what Hamilton had simply called 'latent thought'), was helpful and ingenious, though it recognised that where spiritualist phenomena could not be so explained, Carpenter veered to ridicule. Ridicule, it noted, had also been very effectively applied in Crookes' reply to Carpenter, who was personified as a bombastic opinionated loudmouth who never listened to a different point of view. Even *Chemist & Druggist* felt Carpenter had gone too far in his *ad hominum* denunciations by cruelly saying that Huggins lacked a broad scientific culture, that Cox was a gullible lawyer, and that Crookes had won his scientific spurs (i.e., his FRS) by a lucky fluke. The modern reader, recalling that both Huggins and Crookes later became presidents of the Royal Society, will agree.

Carpenter, an apostle of what historians of Victorian science have called 'scientific naturalism', saw himself as a policeman patrolling science and popular culture for errors and mis-judgements. Thus, when Crookes produced his first article in the *Quarterly Journal of Science* announcing that he intended to investigate the phenomena occurring at séances, Carpenter was on the warpath immediately. It was largely due to his efforts that when Crookes attempted to read a paper on his experiments with Home at the Edinburgh meeting of the British Association in August 1871, it was disallowed – as Spiller had noted. And when Crookes submitted it to be read at the Royal Society, he found the secretaries would not accept it. Crookes therefore published it in his own journal, causing Carpenter to write the anonymous essay 'Spiritualism and its Recent Converts' for the *Quarterly Journal* in October 1871. He claimed in this that the evocation of a 'psychic force' showed that Crookes had become a spiritualist. Crookes always denied this publicly and put up an agnostic front; however, his letters and writings in the spiritualist press do not support his stance.

Provoked by Crookes' sarcastic pamphlet, *Psychic Force and Modern Spiritualism*, Carpenter slandered Crookes in a lecture at the Vestry Hall in Chelsea on 19 January 1872 by publicly revealing confidential information that the Royal Society had turned down Crookes' report of tests on Home. He claimed that he had been told this personally by the Royal Society's Secretary, George Stokes, as well as by Charles Wheatstone. Crookes immediately took up the matter with both men, as well as with the society's president, Sir George Airy, who formally apologised for the leak of confidential information.

Although Carpenter appeared to let the matter drop, some of his remarks in his presidential address to the British Association at Brighton in August 1872, 'On Man as the Interpreter of Nature', were taken by Crookes to be directed at him.[29] Consequently,

[29] W.B. Carpenter, 'On Man as the Interpreter of Nature', *British Association Reports* (1872), lxvix–lxxxiv. See Crookes' review, *Quarterly Journal of Science* (October 1872), **9**, 509–16.

when reviewing the British Association meeting in the October issue of his *Quarterly Journal of Science*, we find Crookes taking pleasure in ridiculing Carpenter's 'childish remarks' on spectroscopy – a field in which Crookes was expert.

The dispute might have rested there; instead vehemence broke out once again when Crookes began to publish reports of physical materialisations in *The Spiritualist* in February and April 1874, while, simultaneously reading a ground-breaking experimental paper to the Royal Society on 'the action of heat on gravitating masses'. All of Crookes' controversial papers on his psychic investigations were also republished at the end of 1874.[30] At first, Carpenter kept quiet, perhaps because Crookes was not challenging scientific orthodoxy by trying to publish in the scientific press. In any case, Crookes appeared to be unveiling amazing new experimental phenomena in physics. Indeed, by the following spring, 1875, Crookes and Gimingham had devised the radiometer bulb, which was again reported to the Royal Society. Simultaneously, however, Crookes reported on the genuineness of psychic phenomena produced by the American medium/magician, Mrs Eva Fay, for whom he had devised an electrical circuit test.[31] Still Carpenter stayed silent.

Instead, on the physical front Crookes found his work on radiation criticised by Osborne Reynolds in Manchester, who challenged his tentative hypothesis that the pith vanes of the radiometer were affected by light. Despite an exciting debate over the physical mechanism causing the radiometer's rotation, everything seemed to be moving in Crookes' favour as the supreme empiricist and experimentalist. In November 1875 Crookes was awarded a royal medal by the Royal Society, a reward for which Carpenter congratulated him and said, 'let's bury the hatchet!'.

Carpenter claimed later that he had been forced to take up the hatchet again because of Crookes' endorsement of the mediumship of Mrs Eva Fay in *The Spiritualist* in March 1875. Although he had ignored this at the time in favour of Crookes' other physical experimentation, he took umbrage at the end of 1876 because he had learned that Mrs Fay and other American mediums were advertising their powers as supported by Crookes 'and other Fellows of the Royal Society'. Carpenter now exploited for all its worth the seeming contrast between Crookes' approach to the interpretation of the workings of the radiometer to that of his interpretation of the wiles of Mrs

---

30 'Notes of an Enquiry into the Phenomena called Spiritual During the Years 1872–1873', *Quarterly Journal of Science* (January 1874), **3**, 77–97; 'Miss Florence Cook's Mediumship', *The Spiritualist* (6 February 1874), **5**, 29; 'Spirit Forms', *The Spiritualist* (3 April 1874), **5**, 157–8; 'The Last of Katie King', *The Spiritualist* (5 June 1874), **5**, 270–71. All of these were republished in Crookes (1874). W. Crookes, 'On the Action of Heat on Gravitating Masses', read 11 December 1873, *Proc. Roy. Soc.* (1873–74), **22**, 37–41.

31 Crookes first exhibited the radiometer at a Royal Society soirée, 7 April 1875. See his 'On Attraction and Repulsion Resulting from Radiation', *Phil. Trans.* (1876), **166**, 325–76. 'A Scientific Examination of Mrs Fay's Mediumship', *The Spiritualist* (12 March 1875), **6**, 126–8.

Fay. A convenient vehicle was James Knowles' new interdisciplinary periodical, *The Nineteenth Century*, a journal specifically designed to discuss intellectual beliefs and doubts as exemplified in the discussions of the Metaphysical Society.

Carpenter's *Nineteenth-Century* article (based on a lecture given at the London Institution in December 1876) began innocuously as a popular, very clear account of the radiometer's history and appearance. He went on to describe how virtually everyone was initially taken in by its apparent working by the impact of light on the surfaces of the vanes until Osborne Reynolds and Arthur Schuster enlightened them in 1876.[32] Now, said Carpenter, following Reynolds' and Schuster's work, the consensus was that radiant energy was responsible and that a mechanism involving the familiar kinetic theory of molecular motion was involved. There was no new force or new principle at work, 'we simply have a known mode of force acting under the peculiar conditions [of vacuum]'. This was confirmed firstly by Crookes' perfection of the vacuum such that the rotation became retarded when too little residual gas remained to show thermal disequilibrium. On the contrary, had light been the cause, the mill would have spun faster and faster. Secondly, it was confirmed because when the air in the bulb was replaced by different residual gases, and exhausted to the same pressure point, the mill rotated at different rates proportional to their specific heats. All in all, Crookes thoroughly deserved the award of the royal medal for confirming the kinetic theory. It seems fortunate that Carpenter was unaware of Crookes' initial assumption that the radiometric effect might be a way of detecting psychic force!

At this point Carpenter plunges in a knife! The moral or lesson of this investigation and discussion of mechanism was the contrast between Crookes' brilliantly systematic investigation of the radiometer and his 'thoroughly unscientific course' of action when investigating spiritualism for which he had invented a new occult psychic force. Crookes had attacked fellow scientists for choosing to ignore Home in 1871, whereas the onus had really been on Crookes to show that Home did not push the board down to register an increased weight on the spring balance. Instead, he had posited a psychic force that could not be commanded at will (i.e., not subject to laboratory repetition). Like the biologist turned spiritualist, Alfred Russel Wallace and the deceased chemist William Gregory, Crookes lacked the ability to discriminate between facts and inferences, and the major need for a higher quality of evidence whenever faced by extraordinary results that appeared to defy common sense. Crookes' investigation of the radiometer had been replicable by others. Crookes had perhaps gone overboard

[32] W.B. Carpenter, 'The Radiometer and its Lessons', *Nineteenth Century* (April 1877), **1**, 242–56. O. Reynolds, 'On the Forces Caused by the Communication of Heat between a Surface and a Gas', *Phil. Trans.* (1876), **166**, 715–24; A. Schuster, 'On the Nature of the Force Producing the Motion of a Body Exposed to Rays of Heat and Light', *Phil. Trans.* (1876), **166**, 715–24; J.C. Maxwell, 'On Stresses in Rarified Gases Arising from Irregularities of Temperature', *Phil. Trans.* (1879), **170**, 231–56.

in inferring that he had developed a new photometer, but he had shown true scientific spirit in continually amending his explanations until the *vera causa* of radiant heat was found. What a contrast, Carpenter observed, to Crookes' January 1874 publication 'Notes of an Inquiry into Phenomena Called Spiritual' which showed Crookes' mind to have a completely unscientific side to it, being possessed by a 'dominant idea'. Crookes had been deluded by the idea of psychic force and by the deceits practised on him. Carpenter ended his examination by drawing parallels with the Salem Witch Trials of the seventeenth century when honest judges and juries had been deluded by the idea of devil possession, as they later admitted.

Despite the interesting polemical issues raised by Carpenter, his conclusion was rather weak and contrived. The duality of Crookes' mind taught a lesson, he deduced, namely the need to train the whole mind in the discipline of logical reasoning. This conclusion was nothing new and really only repeated an accusation Carpenter had made before in the *Quarterly Review* of 1871. On that previous occasion, in his pamphlet reply Crookes had easily refuted any imputation that he had been improperly trained by recounting his multi-disciplinary curriculum vitae. In fact Carpenter had actually missed a trick by failing to notice that the radiometric work had led Crookes to the distinctly speculative conclusion that he had forced matter into a new state with extraordinary powers, the fourth radiant state of matter in the continuum of solid, liquid and gas. Crookes was later to imply that this fourth tenuous state of matter was a possible link between our world with others – the material that Pullman exploits in his wonderful trilogy, *His Dark Materials*.

Crookes' answer to Carpenter's latest uncomfortable *ad hominem* attack was threefold. He commissioned Wallace to write a critical review of a book on spiritualism that Carpenter published in 1877 and, unsuccessfully, asked Wallace to persuade the marine biologist George Wallich, to attack Carpenter's credentials as a marine biologist.[33] He also ably refuted Carpenter's argument in a reply entitled 'Another Lesson from the Radiometer' published in the *Nineteenth Century*. Although its title was logical enough in a rhetorical sense, it was also unfortunate since it seemed to imply acceptance of Carpenter's first lesson.[34] The essay again showed Crookes as a consummate polemicist. His principal point was that writers of popular articles have a duty to be accurate. Carpenter was not. Carpenter was, as historians would say today, Whiggish to exploit the latest interpretation of a phenomenon to criticise earlier ones. In an 'age of research' a critic needed to be an expert in the field. Carpenter was not

---

[33] W.B. Carpenter, *Mesmerism, Spiritualism, &c, Historically Considered*, London, 1877; [A.R. Wallace], 'Review of Carpenter', *Quarterly Journal of Science* (1877), 2nd ser. **7**, 391–416. For the commission and reference to Wallich, see Crookes to Wallace, 24 May 1877, BM Add. 464.39 f.139b.

[34] W. Crookes, 'Another Lesson from the Radiometer', *Nineteenth Century* (July 1877), **1**, 879–87.

a physicist. Crookes claimed he had never been totally committed to the mechanical force of light theory, which was why the instrument had been called a radiometer. Curiously, he pointed out, the term 'light mill' was being used by the very people (e.g., Reynolds) who denied the mechanism was light. When speaking of 'weighing a beam of light' at a Royal Institution lecture he had been speaking figuratively (as he had made explicit at the time); Carpenter had ignored this.[35] Then, it was back to old scores. When Carpenter attempted to show the duality of Crookes' mind he had invoked the experiments with Home; but, said, Crookes, when Carpenter had publicly criticised these experiments at Chelsea Town Hall in 1871 he had misdescribed them and completely misled the audience concerning their rigorous scientific character. Carpenter's criticisms, he concluded, were untrustworthy.

Crookes' own 'lesson' from the affair was more significant than Carpenter's, he believed. It was the importance of following up 'residual phenomena' or anomalies in scientific research. He had in mind his observation of an unusual and unremarked green spectral line in samples of selenocyanide wastes that he had tested; pursuing the line had led to the discovery of thallium. The unexpected repulsive effect of a balance pan when engaged in the determination of thallium's atomic weight by weighing *in vacuo*, had led him to the radiometer, the most talked about scientific instrument of the decade. Such anomalies were noted by 'men of disciplined mind and of finished manipulative skill', he observed – implying that Carpenter lacked such qualities. As for common sense (a point Carpenter had raised in connection with Crookes' psychic claims), that was just what men used to disguise ignorance. After all, what was more bizarre to common sense, spiritualism or the kinetic theory that invisible molecules are dashing around randomly like express trains?

At this juncture, the experimental physicists who had recently formed the Physical Society leaped to Crookes' defence. George Carey Foster, the professor of Physics at University College, London, attacked Carpenter at the Plymouth meeting of the British Association in August 1877, praising the elegance and brilliance of Crookes' sustained investigation of the radiometer.[36] This led to a slanging match between Foster and Carpenter in *Nature*, to which Crookes also made one final witty intervention by sarcastically ridiculing Carpenter's determination to identify him as an allotrope of *ortho*- and *pseudo* Crookes.[37]

Crookes undoubtedly worsted Carpenter in this final act of the debate. However, despite the effectiveness of Crookes' reply to what he called a 'nasty spiteful article' (to

---

[35] W. Crookes, 'Weighing a Beam of Light', *Engineering* (18 February 1876), 133–4.

[36] G.C. Foster, 'Presidential Address', *Nature* (1877), **16**, 311–14.

[37] The debate can be followed under the title 'The Radiometer and its Lessons' in *Nature* (1877), **16**, 544–6 and (1877–78), **17**, 7–9, 26–7, 43–4, 61–2. For an excellent analysis, see Richard J. Noakes, *'Cranks and Visionaries': Science, Spiritualism and Transgression in Victorian Britain* (unpub. PhD thesis, University of Cambridge, 1998), ch. 4.

his son Henry[38]), and the way in which Crookes made the controversy fun to observers through his comic ingenuity for ridicule, privately he was thoroughly shaken by the continuation of Carpenter's vehemence. In October 1877 he was dismayed to hear of Royal Society gossip that several fellows felt that Carpenter had been right not only in his analysis of Crookes' changing views of the radiometric phenomenon, but also in the lesson drawn that Crookes could not be trusted to interpret the facts. Walter White, the administrative secretary of the Royal Society, reported in his diary that Crookes was 'unhappy, can't settle to work, says his self respect will compel him to answer any further attacks or charges'. White recorded that he advised Crookes that 'silence and honest work, work for the love of it, are stepping stones to greatness'.[39] Fournier d'Albe, Crookes' first biographer, was of the view that Carpenter's purpose and triumphal success was to ensure that spiritualism did not invade the doors of the Royal Society. In consequence, like other hobbies and disciplines, it had to become the province of a new institution, the Society for Psychical Research in 1882. Carpenter justified himself by portraying his mission as that of a public health doctor stamping out a mental disease that threatened the rationality of England. To Crookes' evident embarrassment, he had been able to point out that Home himself had exposed most of his contemporary mediums on both sides of the Atlantic as frauds in 1876.[40] To add to Crookes' woes, he had been subjected to threats of scandal by fraudulent mediums.[41]

There was room for further embarrassment because both Crookes and Carpenter sat on the council of the Royal Society in 1877. Carpenter's sniping did not stop, for in November 1877 he fired off a letter to *Nature* deriding Crookes' support for the electrical tests he had performed on Mrs Fay in 1875. Crookes made a bitter answer saying it was his last word: Carpenter was wasting his valuable time better spent in the laboratory.[42]

Only from 1878 did matters calm down between the two antagonists. Carpenter spent the remaining seven years of his life absorbed in oceanography, while Crookes threw himself into wonderful experimental work on the trajectories of molecules in evacuated spaces. He is best remembered for this experimental work that led directly to the identification of the electron by J.J. Thomson. The irony is that his guiding

---

[38] W. Crookes to Henry Crookes, 11 May 1877, in D'Albe, *Crookes*, 263–6, 264.

[39] *The Journals of William White*, London, 1898, 265 (entry for 23 October 1877). Crookes subsequently urged Wallace to ignore Carpenter's attacks on himself, but to no avail. Crookes to Wallace, 6 December 1878, quoted by D'Albe (1923), *Crookes*, 295.

[40] D.D. Home, *Lights and Shadows of Spiritualism*, London, 1877. Curiously, Home included a letter from Crookes (21 January 1876, 183) trying to dissuade him from publishing the book.

[41] Medhurst and Goldney (1964).

[42] W.B. Carpenter, 'Mr Crookes and Eva Fay', *Nature* (8 November 1877), **17**, 81, 122–3; W. Crookes, 'Mr Crookes and Eva Fay', *Nature* (10 January 1878), **17**, 200.

idea was that of radiant matter, a material from which the universe had evolved and a magical stuff that might link our world to others in different dimensions. It was no more, no less than a transformed and materialised psychic force.

There were two significant repercussions of this debate, one positive, and the other negative. In the first place, as Knight has shown in analysing other Victorian scientific debates, scientific rows drive the participants to formulate new ideas and experiments. This is obviously the case with Crookes' superlative investigation of the radiometer's mechanism in which the critical interventions of Reynolds, Schuster, Maxwell, Stokes and many others, provided him with an exciting stimulus.[43] On the other hand, it seems highly likely that Carpenter's repeatedly abrasive offensives on Crookes' investigation of spiritualist phenomena led him to adopt an obstinate attitude towards any self-criticism. Although there are indications that privately he accepted that a large amount of the mediumship that he had witnessed or tested was fraudulent, he would never admit publicly that he had been hoodwinked. He even refused to accept that Eva Fay had got the better of him in the electrical tests he had conducted in 1875, despite clear evidence that her career was based on stage illusions. To admit that Home or Florence Crook had deceived him would have been to admit Carpenter had been right. Given his international celebrity status from the mid-1870s onwards, it was far easier to adopt the same position that his friend Daniel Home had done in *Lights and Shadows of Spiritualism*: fraud existed, but there was a residuum of spiritualist phenomena that resisted scientific scrutiny and formed its own mystery.

The dispute between Carpenter and Crookes reads, on the one hand, like a rehearsal for Karl Popper's argument for fallibility or falsificationism as the criterion for separating science from non-science or nonsense. Crookes' work with Home, Cook and Fay could not be replicated in the laboratory and had, in any case, been conducted under lax conditions. On the other hand, we must recognise that Crookes did not accept that some strange, but natural, phenomena that had undoubtedly been witnessed by him, could be ignored by science simply because replication was elusive, difficult, irregular or impossible. Anomalous events must not be dismissed, but pursued, since they could lead to new knowledge. The synchronicity of Crookes' investigation of mediumship, and of weight measurements at reduced pressures, cannot be separated into non-scientific and scientific activities. Initially, Crookes was inclined to suspect a connection – possibly though the existence of a more sophisticated theory of gravitation.

Although this connection was soon dropped, and the research programmes went their separate ways, the false connection was undoubtedly a spur to the development

---

[43] See A.E. Woodruff, 'William Crookes and the Radiometer', *Isis* (1966), **57**, 168–98; S.G. Brush and C.W.F. Everitt, 'Maxwell, Osborne Reynolds, and the Radiometer', *Hist. Stud. Phys. Sciences* (1969), **1**, 105–25.

of the early radiometer. Crookes was therefore undoubtedly correct in retorting to Carpenter that the real lesson of the radiometer was never to ignore observational and experimental anomalies, however irrational they might appear. Carpenter believed that science encompassed only materials and events that were observable, weighable, measurable, twistable, dissectible and vivisectible; in contrast, Crookes believed in an extended definition of science that included the intangible, the odd, quirky and disagreeable. To Carpenter, Crookes seemed to have a split personality. In chemical language he was both *ortho*(*dox*) and *para*(dox), and in the latter mood he was not to be trusted.

The controversies that developed in the Victorian periodical press over Crookes' concurrent research into the radiometric effect and spiritualism had serious repercussions for his standing in the scientific community. He fought back with all the rhetorical weapons at his command and – to continue the chemical metaphor – he *meta*morphosed into a brilliant lecture-demonstrator who inspired and *meta*grobolised specialised and popular audiences alike. What began as an outstanding Victorian row exposing the metaphysical beliefs of two representatives of the new class of professional men of science and the tensions within scientific naturalism during the 1870s ended with Crookes triumphantly becoming an international scientific celebrity. Like Philip Pullman's sacred glass dome, Crookes' whirling radiometer vanes did teach a moral lesson: the black of scientific ignorance fleeing from the white of a new understanding of fundamental physics

Chapter 15

From Science to the Popularisation of Science: The Career of J. Arthur Thomson

Peter J. Bowler

Among the small number of early twentieth-century scientists who gained a reputation for being able to write for a nonspecialist readership, J. Arthur Thomson was one of the most prolific. As professor of Natural History at Aberdeen he occupied a well-established (if not well-supported) academic position which allowed him to be seen as a spokesman for the scientific community. Despite a heavy teaching load he became active in lecturing outside the university and began publishing textbooks at an early stage in his career. He went on to write an enormous number of articles and books aimed at a broader readership, some of which were aimed at the mass market well beyond the affluent middle classes. He was thus in a position to exert a significant influence on the general public's perception of science and its implications.

Thomson is little-remembered today because he is not associated with any major scientific advance. Indeed, he remained loyal to a neovitalist vision of the life sciences which gradually become outdated within academic biology. He did little research in the later part of his career, and what he did publish at this level was confined to an old-fashioned morphological approach. Many thought he had abandoned research altogether for a career as an educator and popular writer. But in using his books and articles to promote a nonmaterialist interpretation of science which allowed for a synthesis with religion, he was presenting an image of science that would not necessarily have been accepted by other members of the scientific community. In 1933 the Rationalist writer Joseph McCabe singled out Thomson along with C. Lloyd Morgan and Oliver Lodge as examples of senior scientists who were misusing their position to promote an outdated view the current direction of scientific research.[1] According to McCabe, the literature produced by these elderly and unrepresentative figures was seriously distorting the public perception of science, helping to sustain

---

[1] J. McCabe, *The Existence of God*, rev. edn, London, 1933, 77. On the debates over the wider implications of science in this period see Peter J. Bowler, *Reconciling Science and Religion: the Debate in early twentieth-century Britain,* Chicago, 2001.

a false impression that it had turned its back on the materialism of the Victorian era. Whether or not we accept this claim, it does point out the importance of popularisation in shaping the wider perception of science and its implications. Those scientists who do the latest research are not necessarily those who write for a nonspecialist audience, and the latter may thus be popularising a form of knowledge that the active researchers would repudiate.

Thomson's career may help us to understand the complex relationship between science as it is done in the laboratory and as it is perceived by the outside world. It drives another nail in the coffin of what has been called the 'dominant view' of popularisation, the assumption that popularisation is a derivative process in which technical knowledge generated by research scientists is simplified for presentation to a passive audience among the general public. This model has already been much criticised by sociologists and historians.[2] One problem is the complexity of the publics being addressed, ranging from university students to professionals in other fields and the readers of mass-market books and newspapers. Thomson certainly wrote at all these levels and it is of interest to know how he structured his career as a writer to fit them all in. More seriously, though, the dominant model assumes that there is an uncontested version of scientific knowledge accepted by the professional community and available for dissemination to the public. When there are debates within science itself, that is clearly not the case, and Thomson's career as a populariser shows how a scientist can use his popular writing to promote a view of science which would not be accepted by all or even a majority of his colleagues. Some of his writing may have been aimed at persuading other scientists that his views should still be taken seriously. But more often they were intended to persuade outsiders that a particular position was still active – even when in fact it was being abandoned by a majority of younger researchers. This was popularisation used as a weapon in an ideological battle, both inside and outside the scientific community.

To unpack these issues we need to ask why and how scientists become involved with nonspecialist writing, and to explore the distinction between nonspecialist and popular writing. In the Victorian era it was taken for granted that senior figures in science should engage in debate with fellow intellectuals in the pages of the 'highbrow' periodicals. Few scientists, however, ventured to write genuinely popular science – that is, accounts of science and its implications aimed at a readership beyond the educated elite. A few, including T.H. Huxley, did become skilled at lecturing and to some extent writing for a working-class audience. In Huxley's case this was almost

---

[2] E.g. S. Hilgartner, 'The Dominant View of Popularization: Conceptual Problems, Political Uses', *Social Studies of Science* (1990), **20**, 519–39; R. Cooter and S. Pumphrey, 'Separate Spheres and Public Places: Reflections on the history of science popularization and on science in popular culture', *History of Science* (1994), **32**, 232–67.

certainly done because he realised that the nonscientists who were writing most of the popular accounts of scientific discoveries were promoting a modernised form of natural theology opposed to the naturalistic philosophy which Huxley and his associates espoused.[3] Only later in the century did C.A. Watts' Rationalist Press Association begin its campaign to promote a materialist view of science to ordinary people, and then Huxley was unwilling to associate himself too openly with this level of activity.[4] The scientists of this period were willing to engage in intellectual debate, but much less anxious to become involved with the business of purveying knowledge of science to the masses. This left the construction of a wider public image of science to nonscientists who were professional authors, and only at the end of the century did Watts revive the older tradition of radical popularisation which seemed to have gone out of fashion in the mid-Victorian era. The construction of a popular image of science was thus a contested area, yet one which few professional scientists felt comfortable with.

The situation which Thomson encountered in the early twentieth century was significantly different. By this time the scientific community had expanded and had become much more thoroughly professionalised. Much of the expansion had occurred outside the ancient universities and was concerned with the application of science and with technical education. As later left-wing critics alleged, a whole generation emerged for whom it was no longer acceptable for the scientific expert to engage in philosophical or ideological debate. To do so was to risk loosing one's scientific credibility, and we know that this form of suspicion could have serious effects on a career, as when both Julian Huxley and Lancelot Hogben encountered delays in getting elected to the Royal Society.[5] Yet the source and the magnitude of the threat can easily be misunderstood. Huxley was indeed gradually withdrawing from his research career when he was finally elected FRS, and Hogben was elected despite the controversial nature of his public pronouncements. Other eminent scientists continued to enjoy the respect of their peers despite gaining a wide reputation through nonspecialist writing, including J.B.S. Haldane, James Jeans and Arthur Eddington. The trick was to ensure that a substantial contribution to research was maintained at the same time, so there could be no question of seeming to abandon 'real' science for popularisation. It also helped to have an established university position – unlike scientists working in industry or technical education those at the elite universities were still expected

---

[3] See for instance B. Lightman, '"The Voices of Nature": Popularizing Victorian science', in B. Lightman, ed., *Victorian Science in Context*, Chicago, 1997, 187–211.

[4] Lightman, 'Ideology, Evolution, and Late-Victorian Agnostic Popularizers', in J.R. Moore, ed., *History, Humanity and Evolution: Essays for John C. Greene*, Cambridge, 1989, 285–309.

[5] See C.H. Waddington, *The Scientific Attitude*, 2nd edn, London, 1948, 81. On Hogben see G. Wersky, *The Visible College: a Collective biography of British scientists and socialists of the 1930s*, new edn, London, 1988, 165.

to be articulate intellectuals. Indeed, in contrast with the late Victorian era, a large proportion of the nonspecialist writing on science in the early twentieth century was done by working scientists in their spare time. Publishers seem to have preferred their authors to have an established professional position, because this lent authority to their pronouncements. Much of this nonspecialist writing was still aimed at the educated middle classes, of course, but even books intended for this audience could reach large numbers of people if they attracted enough attention in the press, as was the case for Jeans and Eddington's accounts of developments in physics and cosmology. Very few scientists actually gained access to genuine mass-market publishing, writing for the daily newspapers and for book publishers who were prepared to issue cheap, well-advertised series aimed at those who would only ever buy a handful of books in their lifetimes.

Along with Julian Huxley, Thomson was one of those few scientists who gained enough experience with the publishing industry to be asked to write for the mass market as well as for the prosperous and literate middle class. And, like Huxley, Thomson gradually scaled down his career as a research scientist and ended up being marginalised by the scientific community. In his case, there was never to be the coveted FRS, although he gained a knighthood when he retired in 1930. Unlike Huxley, though, Thomson kept his university chair and could be portrayed as a member of the scientific community, even though anyone 'in the know' would realise that he was not considered to be an active researcher. He balanced a heavy teaching load with a crippling schedule of nonspecialist writing and editing, becoming known for his standard (if increasingly dated) textbooks, for his engaging descriptions of natural history and for his robust defence of neovitalism and teleology. It was this last aspect of his work which attracted McCabe's attention, of course, and it is easy to see why he would regard such an effective and apparently authoritative writer as a threat to Rationalism. In the end, though, McCabe's fears were probably misguided. The Rationalist argument was being put before the public with equal skill by scientists such as Arthur Keith and Julian Huxley (who collaborated with H.G. Wells in the hugely successful *Science of Life* of 1931). At best, Thomson was merely holding the fort for natural theology, and when he died in 1933 there was no one to replace him as a popular advocate for this position.

This article explores Thomson's career as a popular writer and propagandist and seeks to understand the relationship between professional science and the way that activity is perceived by outsiders. What drove him to give up an initially promising career as a research scientist for one based largely on teaching and writing? How important a role in this transition was played by his religious beliefs? How did he gain access to publishers in order to maximise his output of nonspecialist writing, and how did he integrate this activity with his academic career as a science professor? And finally (following McCabe's point): how effective was he in reaching out to a wider audience and shaping their vision of science? We cannot answer this latter question solely by looking at reviews of his many books, because these were often written by reviewers defending preconceived positions. We need to look at the practice of writing for a popular audience, at the activities and expectations of authors, editors and publishers as they sought to make money by selling as many books as possible in a competitive market. Perhaps Thomson's deeper involvement in these activities, as compared to most scientist-popularisers, will throw light on how easy it was to promote a particular image of professional science to those who only saw it from the outside.

Thomson's Career in Science

Thomson originally studied at New College Edinburgh, but then moved to work in biology under Ernst Haeckel at Jena. He always remembered Haeckel with affection, although he was disturbed by his reputation as a free thinker.[6] What he did gain from his teacher was an enthusiasm for the great drama of evolution, which Thomson himself would always invest with a religious significance. He returned to Edinburgh in 1886, fully conversant with the latest morphological techniques in biology and the latest ideas about evolution, and began lecturing in medicine and biology at the university. He also began to offer extra-mural lectures to nonspecialist students, the start of his long career as a populariser. In 1899 he was offered the Regius Chair in Natural History at Aberdeen, where he stayed until his retirement in 1930.

Unlike Julian Huxley, Thomson's involvement with popular writing was never deep enough for him to give up his chair. At the start of his career he was perceived as someone capable of making a significant impact on the field. But the amount of time he began to devote to popular writing, coupled with his heavy teaching load, left him little time for original research. By the 1920s his links to the research community were becoming increasingly tenuous, and what little work he did remained within the now outdated methodology of Haeckel's morphology. He himself recognised that his ideas were no longer in tune with the thinking of the younger generation of biologists who had turned to experimentalism and reductionism.

Thomson made some impact on scientific thinking at the start of his career through the publication in 1889 of his first collaboration with his friend Patrick Geddes, *The Evolution of Sex*. Here they made an attempt to understand the role of sex in evolution and the role of evolution in producing sexual differences within species. The book remained in print until at least 1913, when the authors fell out with the publishers

---

[6] For Thomson's recollections of his time with Haeckel see his *The Great Biologists*, London, 1932, 158–61.

over plans to issue a revised edition.[7] By 1896 Thomson had gained a reputation for being extremely well-read, and the editor of the *Zoological Record*, Frank Beddard, asked him to compile a new section covering more general contributions to biology, a task he continued to fulfil until his retirement in 1930.[8] He remained interested in the debates over evolution theory and heredity, and helped his wife to produce the English translation of August Weismann's *The Evolution Theory* in 1904.[9] His own survey of the heredity debates was commissioned by John Murray and appeared in 1907. The book received an enthusiastic review in *Nature* which praised its even-handed approach but also noted its lucid exposition and literary charm.[10] A review of the third edition in 1919 acknowledged that the book had established itself as an introduction 'at once trustworthy, impartial and comprehensive'.[11] Yet there was little original in Thomson's treatment, and a letter to Geddes from an unidentifiable correspondent (the signature has been deliberately obliterated) complained that he was disappointing those who had looked to him as a potential leader in the field by settling for 'smaller things' such as this survey.[12]

By the time of Thomson's death, the assumption that he had effectively abandoned original research was so widespread that it was alluded to in his obituary in *The Times*.[13] This was immediately contested by W.T. Calman, who pointed out that Thomson had continued to publish morphological work on the Alcyonarians (a group of polyps).[14] Calman conceded that this type of research was now disdained by the majority of biologists, who preferred to work on experimental topics. The episode does suggest, however, that Thomson was no longer perceived as an active figure in the key areas of biological research. To the extent that he was still thought to be doing valuable work, it was through his surveys of what others were publishing – a good foundation for a populariser but hardly enough to earn an FRS.

---

[7] Geddes and Thomson, *The Evolution of Sex*, London, 1889; on the revised edition see Walter Scott & Sons to Thomson, 29 December 1913, Geddes papers, Strathclyde University, 9/1236; Thomson to Geddes, 5 January 1914, *ibid.*, 9/1240.

[8] The anonymous obituary of Thomson in *The Times* drew attention to this aspect of his work, 'J. Arthur Thomson: Natural history and religion', *The Times*, 13 February 1933, 8.

[9] A. Weismann, *The Evolution Theory*, trans. J.A. Thomson and M.R. Thomson, London, 1904, 2 vols.

[10] G.A. Reid, 'Heredity', *Nature* (1908), **78**, 361–3, a review of Thomson, *Heredity*, London, 1907. Thomson's preface notes John Murray's encouragement for the project.

[11] GHG, review of *Heredity*, 3rd edn, *Nature*, (1919), **104**, 92.

[12] Letter to Geddes, 1908, Geddes papers, National Library of Scotland, MS 10555, 151.

[13] Anon (1933).

[14] W.T. Calman, 'Sir J. Arthur Thomson', *The Times*, 15 February 1933, 17. The *Nature* obituary by James Ritchie, Thomson's colleague at Aberdeen and successor to his chair, noted his continuing work in this area, 'J. Arthur Thomson', *Nature* (1933), **131**, 1296. See for instance Thomson, et al., *Studies in Alcyonarians and Hydroids*, series 9, Aberdeen, 1909–15.

Lectures and Textbooks

To understand Thomson's increasing involvement with lecturing and writing for nonspecialists, we need to recognise that there were several interacting motivations. One, which we shall return to below, was his concern to ensure that people were aware of his own particular vision of science and its implications. He certainly thought that the public should know more about science, but like many of the best popular science writers he was concerned that they should be introduced to it through a particular interpretation of its significance. He made no effort to pretend that he was promoting a value-free image of the latest discoveries – he was driven to write about science because he thought it ought to mean something to his readers. The problem was that this vision remained constant throughout his career, while the scientific community changed radically through the early decades of the new century, leaving Thomson to some extent isolated from the views of the new generation.

In addition to this ideological motivation, there is no doubt that Thomson did feel that it was his duty to inform as wide a public as possible about developments in science. If the majority of scientists found it hard to write about their work in terms that laypersons could understand, it was up to the few who did have this skill to devote a significant amount of time to this activity – even if it left them vulnerable to sneers from their more research-active colleagues. And there is no doubt that Thomson did acquire the skill of speaking and writing about science in ways that were both intelligible and inspiring to ordinary people. Beginning with extra-mural lectures, he was soon being asked to write these up for publication, and by the start of his career at Aberdeen he was already deeply involved with the publishing industry. He not only wrote for a variety of publishers who seemed only too anxious to commission his work once they realised that he had the necessary communications skills, he also became editor of a popular science series. In this way Thomson built up contacts with the publishing industry that would shape the whole of his career in a direction that was quite untypical.

Even before he began teaching at Edinburgh university, Thomson had become active in the university extension movement, returning in the summers to give extra-mural lectures. In 1885 he complained to Patrick Gedde that he was spending far too much time on this kind of activity.[15] In 1888 he lectured on zoology (the course also included laboratory work and a marine excursion) and gave a course on geographical distribution to the Royal Scottish Geographical Society.[16] But the lectures were merely a starting point, because soon Thomson was being asked to write them up as articles for

[15] Thomson to Geddes, 1 September 1885, Geddes papers, National Library of Scotland, MS 10555, 55–8.

[16] Printed notice of lecture series by both Geddes (botany) and Thomson (zoology) for the 1888 summer season, Patrick Geddes papers, Stratclyde University, 12/2/17; Thomson's syllabus for lecture series at the Royal Scottish Geographical Society, *ibid.*, 13/2/24.

the press or in the form of nonspecialist textbooks. His *Study of Animal Life* of 1892 received a favourable review from Lloyd Morgan, praising its 'exceedingly happy' style of writing, but cautioning that its use of terms such as 'love' to denote the sexual instincts of animals might be misleading.[17] This latter comment illustrates an important link between Thomson's wider beliefs and his success as a popular writer on natural history. Where Morgan avoided attributing the higher feelings to animals, Thomson was only too happy to describe the mental life of animals in anthropomorphic terms. Even when not explicitly endorsing vitalism, he was portraying animals in terms which encouraged the nonspecialist reader to see them as creatures driven by mental powers which transcended mechanistic explanation. Whatever Morgan's reservations, *The Study of Animal Life* remained in print for over a quarter of a century.[18]

Thomson also wrote a detailed textbook for the zoology student, his *Outlines of Zoology*, first published in 1892. This got a very mixed review in *Nature*, the reviewer praising the writing style but complaining about theories being presented before facts and a lack of individuality which left the reader to select what was important.[19] Nevertheless, the reviewer thought the book had real potential, and so it did, going through nine editions, the last published posthumously in 1944. Commenting on the eighth edition (1929) *Nature* noted that in the uncertain market for textbooks, Thomson's *Outlines* had satisfied a real demand and enjoyed undiminished popularity throughout the country.[20]

Successful teaching, both within and without the walls of the university, thus formed a natural stepping stone by which Thomson was led to write for a wider readership than that of the technical scientific article or monograph. He continued to lecture to nonspecialists, some of his popular books (discussed below) having their origins in such series. By the turn of the century it had become clear to everyone, including the publishers, that Thomson had the knack of being able to write about nature and about science at a level which would attract ordinary readers. Demand built up for him to write and edit nonspecialist texts similar to his *Study of Animal Life*, or more popular articles and books of the kind that would attract the casual reader. His career as a populariser thus emerged gradually out of his work as a teacher and lecturer.

Thomson's Career in Publishing

From an early stage in his career Thomson became more heavily involved with publishing than was normal for working scientists at the time. He became an editor as

[17] Lloyd Morgan, 'The Study of Animal Life', *Nature* (1892), **47**, 2–3.
[18] See the preface to the London, 1916 and London, 1923 editions.
[19] G.B.H., 'A Treatise on Zoology', *Nature* (1892), **46**, 241–2.
[20] D.L.M., review of *Outlines of Zoology*, *Nature* (1930), **125**, 269.

well as an author, and thus gained a more intimate understanding of what publishers were looking for. This was especially important when he came to work for firms that were attempting to tap into the demand for genuine mass-market books. He became aware of the perils facing commercial publishers in a competitive and ever-changing market – we shall see that several of the firms he was involved with got into difficulties. At the same time he made contact with the editors of magazines and also of mass-circulation newspapers, and was soon writing regularly for the latter (a very specialised skill because deadlines were rigid and articles often had to be limited to 500 words). This was a cut-throat business which few scientists could accommodate themselves to. Yet it also had its rewards, and Thomson was soon making a comfortable addition to his salary from royalties and fees – yet another incentive to keep up this side of his work.

In the late 1890s Geddes had set up his own publishing company, for which Thomson worked as an editor.[21] He solicited manuscripts from other scientists for a series of relatively cheaply produced books on a range of scientific topics. By 1910 Thomson had become the scientific editor of the Home University Library, a series introduced by Williams and Norgate to satisfy the demands of ordinary people in polytechnics, working mens' colleges, evening schools and literary societies. It was hoped to wean away those now addicted to 'trashy forms of cheap literature' by offering them informative but attractive reading at a price they could afford. Each volume was to be 50,000 words in length and sold at a price of one shilling per copy. The series was heavily advertised and each volume was expected to sell in the tens of thousands of copies. Authors got a penny a copy royalties, with an advance of £50.[22] This meant that an auther could easily earn £100 for a book, at a time when even professors were lucky to earn £1,000 *per annum*, and lecturers far less than this. By the late 1920s the series contained well over 100 titles, of which 34 were in the area of science.[23] Despite its apparent success, Williams and Norgate were by this time in financial difficulties and the series was sold off to another publisher.[24]

This position as editor allowed Thomson to influence the kind of authors who were chosen to contribute. He and Geddes published a number of their own works in the

---

[21] Thomson to Patrick Geddes and Colleagues, 20 and 22 August 1896, Patrick Geddes papers, National Library of Scotland, MS 10555, 80–82.

[22] Details of how the series operated can be seen in the contract between Geddes and Thomson and the publisher for their 1911 book on *Evolution*, Geddes papers, Strathclyde University, 10/1/39, and in the printed instructions for authors, Geddes papers, National Library of Scotland, MS 10555, 153B. On potential sales figures see Thomson to Geddes, 4 October 1910, National Library of Scotland, MS 10555, 154.

[23] List printed at the end of some volumes, this figure from Geddes and Thomson, *Biology*, London, reprinted 1929.

[24] See Thomson's letters to Geddes, 28 November and 11 December 1927, Geddes papers, National Library of Scotland, MS 10555, 290, 292.

series, providing ample opportunity to promote their anti-mechanist view of biology. In 1911 the series published Thomson's *Introduction to Science*, which included chapters endorsing the neovitalist position and the possibility of a new relationship between science and religion. The same year saw the appearance of a volume on *Evolution* which he co-authored with Geddes, the introduction to which makes an interesting comment on the public's thirst for synthetic works at all levels from the learned journals down to *Tit-Bits*.[25] Not surprisingly, the text endorsed neovitalism and the concept of creative evolution, as did their co-authored *Biology* text, published in the series in 1925. Yet Thomson's powers were not absolute – he encouraged Geddes to write a book on 'Cities' for the series (town planning was becoming increasingly central to the latter's interests), but this was eventually vetoed by the other editors.[26]

Editing the Home University Library gave Thomson a level of contact with the publishing industry that was far more intimate than that enjoyed by most scientists who wanted to try their hands at popular writing. But once he had gained a reputation for being able to write at this level, other publishers were constantly pressing him to write or edit for them. Following their success with H.G. Wells' *Outline of History*, George Newnes decided to issue an *Outline of Science* in 1922 – the books were produced in identical formats and were clearly intended to sell to the same very broad readership. Thomson, who admired Wells' book although he was not in sympathy with its Rationalist message, took over the editorship from Joseph McCabe. He included a chapter by Oliver Lodge on spiritualism, which provoked a public rebuke by Lodge's longstanding opponent on this issue, E. Ray Lankester.[27] A three-volume *New Natural History* written by Thomson subsequently appeared in the same format.[28]

Other popular surveys written by Thomson include his *Everyday Biology* of 1923, *Modern Science* of 1929, *Life: Outlines of General Biology*, another collaboration with Geddes, published in 1931 (of which more below) and the posthumously published *Biology for Everyman*. *Everyday Biology* was issued in Hodder and Stoughton's People's Library series and was widely praised. *Nature* 'heartily recommend[ed] the book to the layman who would know something of present-day biology'. The *Glasgow Citizen* praised Thomson's 'happy knack of making a scientific subject clear to the lay mind', while *John o'London's Weekly* noted that 'Professor Thomson has a genius

---

[25] Thomson, *Introduction to Science*, London, 1911; Geddes and Thomson, *Evolution*, London, 1911. Thomson had originally hoped to get E. Ray Lankester to write the book, see his letter to Geddes, 4 November 1910, Geddes papers, Strathclyde University, 1/1/3.

[26] See Thomson's apologetic letter to Geddes, 18 March 1912, Geddes papers, National Library of Scotland, MS 10555, 163–5.

[27] Thomson, ed., *The Outline of Science*, London, 1922, 2 vols; Lodge's 'Psychic Science' is vol. II, ch. XVI. For Lankester's complaint see J. Lester, *E. Ray Lankester and the Making of Modern British Biology*, British Society for the History of Science monographs, 1995, 213.

[28] Thomson, *New Natural History*, London, n.d., 3 vols.

for making science interesting'.[29] *Biology for Everyman* was commissioned by J.M. Dent (founder of the Everyman's Library) as a nonspecialist survey and according to E.J. Holmyard, the editor who prepared the text for publication after Thomson's death, he regarded it as his *magnum opus*. The *School Science Review* hailed the sheer amount of material that had been included in the two volumes for the very reasonable price of 15 shillings and insisted that Thomson's name guaranteed its style and accuracy. It advised its readers to 'purchase immediately for their own use and also for the school library'.[30]

Thomson gained a reputation for being able to write about nature in a way that both entertained and informed the general reader. His massive *New Natural History* tapped into what he believed was a growing interest in this area. He stressed the new insights of ecology, but in a non-technical way which encouraged people to take an interst in the 'haunts of life'. His introduction to these volumes also made it clear that he was determined to depict animals as 'personalities of a sort' who could control their own lives and their species' future evolution.[31] This anthropomorphism was a direct product of Thomson's neovitalist philosophy, but it also made for effective nonspecialist communication. In addition to relatively formal surveys of natural history and general biology, Thomson also wrote popular essays of a more lively and spontaneous nature. These often began as public lectures and were then written up for publication in various magazines and newspapers (which would routinely pay several pounds for even a short article, offering a comfortable addition to his earnings). Eventually Thomson began writing on a weekly basis for the *Glasgow Herald*, perhaps Scotland's leading newspaper at the time, paralleling the tradition established by E. Ray Lankester's weekly 'Science from an Easy Chair' articles for the *Daily Telegraph*.

The first of the books generated by compiling these essays was *Secrets of Animal Life* of 1919, based on lectures originally given at the university and written up for the *New Statesman*. *The Control of Life* (1921) stressed the increasing intelligence of the higher animals and the greater freedom this gave them to control their lives. In 1924 *Science Old and New* brought together articles from the *Glasgow Herald*, *John o'London's Weekly*, *Time and Tide* and the *Illustrated London News*. All of these books were published by Andrew Melrose, a personal friend of Thomson's. Indeed they were close enough that Thomson invested money in Melrose's company, and lost heavily when it went bankrupt in the mid-1920s.[32] Melrose had told Thomson

---

[29] Review of Thomson, *Everyday Biology*, *Nature* (1924), **113**, 780; newspaper quotations from the advertisement at end of Thomson, ed., *Ways of Living*, London, 1926.

[30] E.J. Homyard, preface to Thomson, *Biology for Everyman*, London, 1934, 2 vols, I, vii; review in *School Science Review* (1935), **16**, 575.

[31] Thomson (n.d.) I, Introduction.

[32] Thomson to Geddes, 11 December 1927, Geddes papers, National Library of Scotland, MS 10555, 291.

that to stay in business a publisher either had to specialise, or to have so wide a range of publications that a failure by one could be balanced by probable successes elsewhere.[33] Presumably he had tried the latter tactic and failed – his books would have been published at a price which made massive sales unlikely even when authored by a popular figure such as Thomson. Hodder and Stoughton published Thomson's most openly ecological collection, based on lectures given to the Aberdeen branch of the Workers' Educational Association, his *Ways of Living* (1926).

A more explicit link between Thomson's ideas on animal evolution and human origins had been published as *What is Man?* in 1923. This gained a very positive review from F.S. Marvin, praising the writing style but also noting Thomson's willingness to relate his work to the broader questions raised by science, in this case promoting an optimistic image of evolution as a foundation for human progress.[34] This combination of neovitalism and teleological evolutionism had by now become as much Thomson's trademark as was his effective writing style. His popular lectures, articles and books were all written for a deeper purpose, and to understand that purpose we need to look at his more serious efforts to articulate his philosophy to the educated classes.

Opposition to Materialism

From letters written to Geddes when he was studying at Jena, we know that at first Thomson had doubts about his suitability for a scientific career. He longed to return to New College and study to become a pastor in the Free Church of Scotland.[35] By 1886, however, he had come more strongly under Geddes' influence and was led toward a more flexible vision of nature as the scene of divine activity (he compared his conversion to this view with his original commitment to Christianity).[36] From this point onward he was a passionate advocate of neo-vitalist physiology and a teleological evolutionism which treated the higher qualities of humankind as the

---

[33] Thomson to Geddes, 4 October 1910, Geddes papers, National Library of Scotland, MS 10555, 154.

[34] F.S. Marvin, 'The Making and Ministry of Man', *Nature* (1924), **121**, 266, a review of Thomson, *What is Man?*, London, 1923.

[35] Thomson to Geddes, 2 January, 24 March and 4 December 1883; Patrick Geddes papers, National Library of Scotland, MS 10555/ 1–2, 4–5 and 9–12. I have discussed this aspect of Thomson's career in more detail in an article entitled 'Emergent Evolutionism, Neovitalism and the Reconciliation of Science and Religion', to be published in a volume edited by John Hedley Brooke.

[36] Thomson to Geddes, 10 August 1886; *ibid.*, 48–9. At this point Geddes was a botanist, but he would go on to make his reputation in environmentalism and sociology; see for instance P. Kitchen, *A Most Unsettling Person: an Introduction to the ideas and life of Patrick Geddes*, London, 1975.

intended outcome of universal progress. Several of his works on this theme were written in conjunction with Geddes, but the other major influence on his thinking was the creative evolutionism of Henri Bergson. Thomson wrote a very positive review of *Creative Evolution* for *Nature*, arguing that although it was nature poetry rather than biology, it was an important complement to science.[37] Like a number of other biologists at the start of the new century, he hoped that Bergson's views along with the anti-mechanist philosophy of Hans Driesch would revolutionise their science. This was more holism or organicism than outright vitalism, although the distinction was often hard to maintain.

Almost from the start Thomson's nonspecialist writing was being used to promote his stand against materialism. In 1904 he contributed a joint essay with Geddes to a collected volume edited by an Anglican clergyman in which they insisted that life was something more than a mechanical phenomenon and that evolution had a moral purpose.[38] In his Gifford lectures for 1915 and 1916, published as *The System of Animate Nature*, Thomson noted both Hans Driesch and J.S. Haldane's critiques of mechanism, stressing the need for a methodological vitalism which used non-physical characteristics to describe living things without implying that they were driven by vital forces.[39] His *Science and Religion* also stressed that life had to be seen as a creative force that could not be accounted for by the laws of the physical world.[40] In 1925 Geddes and Thomson's *Biology* stressed the role of non-physical explanatory systems so heavily that they conceded that some readers might think they hankered after Driesch's entelechy or Bergson's *élan vital*. Their real purpose, though, was to argue that the psychological characters of living things gave them a degree of spontaneity – this was the real life force.[41] A few years later the last great collaborative work by the two men openly proclaimed itself the vehicle for projecting a neovitalism which showed 'the untenability of a Biology which denies that Mind counts'.[42]

Thomson's teleological evolutionism was proclaimed in his Gifford Lectures, where he insisted that evolution exhibited a general trend toward progress, although of course there were side-branches where stagnation occurred. The human species was 'the summit of the whole' because we exhibit the greatest ability to control our own

---

[37] Thomson, 'Biological Philosophy', *Nature* (1911) **87**, 475–7.

[38] Thomson and Geddes, 'A Biological Approach', in J.E. Hand, ed., *Ideals of Science and Faith*, London, 1904, 49–80.

[39] Thomson, *The System of Animate Nature: the Gifford Lectures delivered in the University of St. Andrews in the Years 1915 and 1916*, London, 1920, chs 4 and 5.

[40] Thomson, *Science and Religion*, London, 1925, 100–102.

[41] Geddes and Thomson, *Biology*, London, 1925, 172–4, 273.

[42] Thomson and Geddes, *Life: Outlines of general biology*, London, 1931, 2 vols, I, vii. On the writing of this book see below.

behaviour: we are the outcome of a persistent trend toward freedom of mind.[43] He believed that the wonders of nature revealed a creator who could be worshipped.[44] His *Science and Religion*, based on lectures given at the New York Theological Seminary, used the idea of progressive evolution to undermine the traditional notion of a state of inevitable conflict between the two areas. He argued that although they addressed different questions, they had to interact over the significance of the material universe.[45] The real problem was that the minds of living things had all too often been ignored by evolutionists. His own viewpoint was that of panpsychism – although he praised the concept of emergent evolution, at heart he did not believe that mind emerged from matter. It must have been there within nature all the time – and behind it the will of God.[46] The unity of nature, the self-directing activity of life and the progress of evolution toward moral beings do not actually prove the existence of God, but they are all suggestive.[47] Other books by Thomson developing the theme of mental and moral progress in evolution included *The Control of Life* (1921), *What is Man?* (1923), *Concerning Evolution* (1925) and the posthumously published *Biology for Everyman* (1934).

Thomson was also an indefatigable author of articles for collaborative efforts to promote a new image of science. In 1923 he contributed an article on Darwin's influence to F.S. Marvin's *Science and Civilization* stressing 'the age-long man-ward adventure that had crowned the evolutionary process upon the earth'.[48] In 1928 he wrote a piece entitled 'Why we Must be Evolutionists' for Frances Mason's *Creation by Evolution*.[49] He provided the introduction to the follow-up volume, *The Great Design*, proclaiming openly that nature displayed its origins in a creator who wanted it to evolve toward the highest values recognised by humanity.[50] Thomson repeated these ideas in a BBC broadcast in 1931 and in his *Gospel of Evolution*, which again insisted

---

[43] *Ibid.*, 397, 565–6. The claim that there was a central 'trunk' to the tree of life was not uncommon at the time, see Peter J. Bowler, *Life's Splendid Drama: Evolutionary biology and the reconstruction of life's ancestry, 1860–1940*, Chicago, 1996, 424–35.

[44] *Ibid.*, 651–2.

[45] Thomson (1925), 2 and ch. 1 generally.

[46] *Ibid.*, 43–4, Thomson's italics.

[47] *Ibid.*, 177.

[48] Thomson, 'The Influence of Darwinism on Thought and Life', in S.F. Marvin, ed., *Science and Civilization*, Oxford, 1923, 203–20, see 217.

[49] Thomson, 'Why we Must be Evolutionists', in F. Mason, ed., *Creation by Evolution*, New York, 1928, 13–23. Thomson and Lloyd Morgan corresponded about the efforts of 'the dear and energetic Mrs Mason' to promote the new natural theology; this comment is from Thomson's letter to Lloyd Morgan, 4 May 1930, which also notes the success of her books; Morgan papers, Bristol University Library, DM 128/432.

[50] Thomson, 'Introduction', in. F. Mason, ed., *The Great Design: Order and progress in nature*, London, 1934, 11–16, see 14.

that nature is a 'materialized ethical process' and that '*Nature became articulate and self-conscious in Man*'.[51]

Thomson thus became widely known as a scientist who was both skilful at putting his material across to the nonspecialist reader, and a tireless advocate for the nonmaterialist approach to the life sciences. Along with Oliver Lodge and Lloyd Morgan, his was one of the names most frequently quoted by clergymen seeking reassurance that science really had turned its back on Victorian naturalism and was reader for a synthesis with liberal religion. The physician and poet Ronald Cambell MacFie dedicated one of his vitalist tracts and a poem to Thomson in honour of his efforts to promote this vision.[52] Given Thomson's early religious experiences and the enthusiasm with which he welcomed Geddes' less formal approach to spiritual experience, it seems hardly coincidental that much of his output should seek to put this message across to his readers.

The effort Thomson devoted to promoting this message to nonspecialist readers was considerable, and the resulting tensions can be seen in his last major collaboration with Geddes. From the start they wanted to write something more elaborate than the Home University Library format would allow. By 1921 they were planning to spend the summer together in order to write, although Thomson was already warning that the project would take years rather than months.[53] It was not until 1925 that they got a contract from Williams and Norgate for a book provisionally titled 'Essentials of Biology'.[54] Geddes sent Thomson long summaries of the topics he wanted to include, going far beyond Thomson's plans for a 'textbook' approach and calling for extensive discussion of psychology, sociology and the symbolism associated with the life sciences.[55] Thomson agreed to incorporate this material, even though some of it must have seemed very odd even to him. By this time Geddes had lost all contact with working biologists and seems to have felt that he could transform the life sciences by sheer conceptual innovation, while Thomson still hoped to retain some contact with the world of biologists working in the laboratory. The manuscript was not delivered on time, and over the next several years they struggled to complete it. In late 1929 the publishers demanded that the manuscript be finished and by May 1930 the proofreading was underway, the title now having been changed to the one under which it would soon be published.[56]

---

[51] Thomson in Julian Huxley, et al., *Science and Religion: a Symposium*, London, 1931, 23–36; Thomson, *The Gospel of Evolution*, London, n.d., 211 and 203, Thomson's italics.

[52] R.C. MacFie, *Heredity, Evolution and Vitalism*, Bristol, 1912 and 'Simple Beauty and Naught Else', in *The Complete Poems of Ronald Campbell MacFie*, London, 1937, 330–46. On MacFie's work see Bowler (2001), 385–6.

[53] Thomson to Geddes, 23 February and 9 June 1921, Strathclyde University 18/2/43/1 and 2.

[54] The contract, dated 9 February 1925, is preserved at Strathclyde University, 17/11.

[55] Carbon copies of these typescripts may be found in the National Library of Scotland, MS 10555, 224–35 and 238–44.

[56] Thomson to Geddes, 11 October 1929 and 30 May 1930, *ibid.*, 297 and 304.

In June 1931 the book was published, advertised by a brochure outlining its encyclopaedic coverage and stressing that it offered a synthesis transcending the division between mechanist and vitalist theories.[57] Reaction among the scientific community was muted, however, in comparison to the praise lavished on some of Thomson's less openly vitalistic surveys. The book got a positive review in *Nature*, praising its efforts to link neovitalism with experimental biology, but this was by James Ritchie, Thomson's colleague and successor at Aberdeen.[58] In fact Thomson and Geddes knew that their ideas were no longer fashionable among the scientific community. In 1925 Geddes had written to Thomson noting that their whole approach was out of touch with the biology practised at Edinburgh and Cambridge. He knew that Thomson had a loyal public who appreciated his writings, but these were people with no links to academic biology.[59] Geddes thus conceded the point of McCabe's criticism as noted above: by 1930 neovitalism was still being put before the public, but it did not correspond with the ideas of the biologists at the key research centres. Thomson and Geddes knew they were not reflecting the latest biological thinking – they were promoting a vision of nature which fitted their philosophical and religious beliefs, a vision which had inspired an earlier generation of scientists but which was now being marginalised by a mechanistic, experimental approach.

Conclusion

Thomson's career illustrates the complex relationship between science and popular science writing. His textbooks and some of his more formal surveys may fit the classic picture of laboratory-generated knowledge being disseminated to a passive public, but even here he was often dealing with issues (as in the case of heredity) which were still actively debated within the scientific community. Thomson gained a reputation for even-handedness, as well as for the clarity and attractiveness of his exposition. Yet as he began to devote ever more time to popular writing, and less to research, it seems evident that the reputation for even-handedness must have become suspect. He became known as an advocate of a particular philosophy of life which was taken up with enthusiasm only by those seeking actively for a reconciliation between science and religion. Even his popular accounts of natural history served this end indirectly, in the sense that they promoted a view of animals as active psychological agents, thus encouraging his readers to see evolution as a purposeful process and the human spirit as a natural outgrowth of the rich mental life of animals. In the first decade of

---

[57] The brochure is preserved at Strathclyde University, 17/10.

[58] J. Ritchie, 'Science and Philosophy of Life', *Nature* (1931), **128**, 1056–7.

[59] Geddes to Thomson, carbon copy dated 8 February 1926, National Library of Scotland, MS 10555, 238–44

the twentieth century this philosophy could still be presented as a valid contender within biology's current theoretical debates, but by the later 1920s it was obvious even to Thomson and Geddes that the field was moving away from their position. To the extent that Thomson was still publishing research, this was old-fashioned morphological work which did not even reflect his interest in neovitalism. He was popularising a particular vision of science which did not derive from any research activity of his own, and was increasingly out of touch with what those who were research-active were thinking.

This was substance of the complaint made by Joseph McCabe, and while he clearly had his own Rationalist axe to grind it is difficult with hindsight to fault his evaluation. The fact that Thomson and Geddes acknowledged (at least in their private correspondence) a gulf between what they were teaching and what was fashionable at major research centres suggests that they too were aware of the situation. Yet they continued to write, and what annoyed McCabe was the possibility that ordinary readers, not conversant with the shifting priorities of the scientific community, might take Thomson's work as reflecting the dominant view of that community. After all, he was a professor at an ancient (if geographically remote) university and gave the appearance of being in touch with current developments. And as a senior scientists with an unrivalled skill in popular writing, and easy access to publishers built up over years of interacting with them, he was in a position to create a false impression through the sheer numbers of books and articles flowing from his pen. Without his efforts, the amount of apparently authoritative material that could have been used by those seeking to argue that science was still moving away from materialism would have been significantly less. Clergymen seem to have been only too willing to take Thomson's views as typical of a general movement in science. The wider reading public may have been more sceptical, however, because here his mass-market publications were in competition with equally competent and equally well-marketed books by writers such as Wells and Huxley who were promoting the more Rationalist view of science. At best, Thomson helped to keep hope alive for a decade or two among those who were already predisposed to accept his message. Even his vast output could not overwhelm the efforts of those promoting a rival view of science, and once he was removed from the scene (he died in 1933) there was no one to replace him.

Finally, what of Thomson himself? He took a conscious decision to focus on teaching and nonspecialist writing instead of furthering an apparently promising research career. It was not an easy road to follow, as his endless complaints about the shortage of time for serious writing proclaim. But once on that road it would have been difficult for him to retrace his steps, and he seems to have felt that it was his duty to use his skills to promote interest in his field. While recognising that his views were increasingly unfashionable, he felt no discomfort in presenting them to the public as a valid image of science. Like many who get left behind by changing fashions, he must have been convinced that it was everyone else who had taken a wrong turn, so it was still legitimate for him to present the rival alternative. He was never openly repudiated by the scientific community – there were, in any case, many members of his generation still within the profession and still suspicious of the latest trends that were dominating the most active research centres. And everyone at least conceded that he did a good job in arousing public interest in science.

CONCLUSION

Chapter 16

Concluding Reflections

John H. Brooke

The subtitle of this book, 'from natural philosophy to natural science 1700–1900', implies the existence of a major shift in scientific culture – a shift finally discernible by the middle of the nineteenth century, by which time highly specialised enquiries into the workings of nature were less likely to be seen as a branch of philosophy and more likely to be accorded an integrity of their own. This transformation, which in Europe spanned a period of some 200 years, has been characterised in many ways. One of the more influential narratives has been that of the Yale historian Frank Turner who has contrasted the mid-Victorian sense of 'conflict' between science and religion with earlier sensibilities when both natural philosophy and natural history had been associated with a style of theological discourse that purported to unite scientific with religious interests.[1] Turner has argued that the natural theology that had been so pervasive a feature of Anglophonic scientific culture since the late seventeenth century had flourished in the minds of amateur (often clerical) naturalists but finally withered as new standards of rigour were introduced by a generation of campaigners, epitomised by Thomas Henry Huxley, who self-consciously wished to turn scientific expertise into a profession. On Turner's reading, part of the appeal of Charles Darwin's theory of evolution for Huxley was that it contributed to the articulation of a naturalistic world-view that would both embarrass and exclude the clerical amateur. Towards the end of the period routinely known as the 'scientific revolution', it had been possible for Newton to say that discourse about God *belonged* to natural philosophy, whose object was to deduce causes from their effects until one ascended to a first cause that was certainly not mechanical. By the 1860s and 1870s, Darwinism was fuelling agnosticism and Huxley would be declaring that extinguished theologians lay around the cradle of every new science like strangled snakes beside the body of Hercules.

Having been asked to reflect on the preceding chapters, it is clear that they have both enriched and qualified this familiar story. There is enrichment from the essay by Barry Gower who has given fresh content to the assertion that in seventeenth-century natural philosophy there was a wider frame of reference than would later be considered appropriate in technical scientific texts. In particular he describes

---

[1] F. Turner, 'The Victorian Conflict between Science and Religion: a Professional dimension', *Isis* (1978), **69**, 356–76.

the quest for an Aristotelian 'first philosophy', a metaphysics that would confer legitimacy and authority on one's repertoire of explanatory concepts. It is indicative of how fundamental this first philosophy was deemed to be that the Greek word for 'theology' was sometimes used as a substitute. Gower suggests that Newton makes an excellent case study in this respect – that he was still seeking to meet an Aristotelian requirement, tacitly acknowledging that his mathematical principles of natural philosophy were themselves in need of underpinning. The suggestion ties in with one of the thrusts of recent scholarship, which has been to broaden the parameters of Newton's natural philosophy and to reinstate a foundational importance for his alchemical and biblical studies. This is not to deny that Newton issued warnings about the conflation of religious matters with the mathematical analysis of nature, but it is to affirm, as did Frank Manuel, that Newton found it extremely difficult in practice to separate completely the interests that he found most consuming.[2] Aspects of his natural philosophy, such as his correlation of space with divine omnipresence, the agency of his gravitational force (which he was never to reduce to an innate property of matter), and the mechanisms he proposed for the re-stabilising of the solar system reflected his preoccupation with how God might act in the world. It is, however, important to recognise that Newton's confidence in divine providence was derived from what he saw as the fulfilment of biblical prophecies and not merely from the organisation of nature.[3]

Central to the narrative expounded by Frank Turner is the concept of professionalisation. The mid-nineteenth-century *conflict* between religion and science has to be understood in sociological and not merely philosophical terms. The clergy were being marginalised even in those societies such as the British Association for the Advancement of Science, in which they had initially been well represented. Questions have been asked about so direct an association of 'professional' science with the demise of natural theology. Among mid-nineteenth-century physical scientists whose work personifies rigorous standards (Michael Faraday, James Clerk Maxwell, William Thomson) one finds the still firm conviction that the order of nature, presented by their science, is revealing of an ultimate creator.[4] Conversely, it is not difficult to find anti-religious sentiments and an antipathy towards physico-theology among scientific amateurs. The essay by Randolph Cock adds another twist to this debate by exposing additional problems with the concept of 'professional' science. He stresses that it was partly because an interest in the sciences had been a prerogative of the gentleman-

---

[2] F. Manuel, *The Religion of Isaac Newton*, Oxford, 1974, 27–49.
[3] J. Force, 'Newton's "Sleeping Argument" and the Newtonian Synthesis of Science and Religion', in N. Thrower, ed., *Standing on the Shoulders of Giants*, Berkeley and Los Angeles, 1990, 109–27.
[4] C. Smith, *The Science of Energy: a Cultural history of energy physics in Victorian Britain*, London, 1998, 239–41.

amateur that naval officers could find it a fitting pursuit. At the same time, the Royal Navy could be said to have facilitated forms of 'professional' science because it was one of few institutions within which science could become a career. Civilian naturalists were paid by the Navy, as were ship surgeons and assistant surgeons. Some officers became specialists in sciences such as meteorology, geomagnetism and the study of tides. The archetypal 'professionaliser' of science, T.H. Huxley, conformed to a pattern in which 'the professionalisation of science was able to hitch a ride on the back of the professionalisation of naval medicine'. Cock is not averse to using the term professionalisation, carefully showing how the Navy could be said to have promoted it in the several senses identified by Jack Morrell. Yet the various opportunities for a scientific career in the Navy blur the distinction between the professional salaried scientist in his laboratory and the country parson indulging an amateur curiosity in botany.

Other chapters in this collection have added their own substance and refinement to the master-narrative. Because arguments for design in nature are to be found in seventeenth-century natural philosophy and natural history (think of Robert Boyle and John Ray), it would be easy to gloss over distinctions between these domains of enquiry if one were only interested in their theological possibilities. We know that Boyle saw more to marvel at in a microscopic mite than in the abstract order of the solar system, and that the converse appears to be true of Newton. But are there other lines to be drawn between natural philosophy and natural history? Harriet Knight's account of Boyle and the machinations of his editors provides an important insight into the epistemological connotations of these terms. In Francis Bacon's account of inductive methodology, it is the arrangement and systemising of the 'facts' of natural history that leads to secure conclusions in natural philosophy – a progression mirrored in the efforts of Richard Boulton and Peter Shaw to impose an artificial order on Boyle's essays that would elevate them from inchoate natural history to a coherent and unified natural philosophy. If this was a mirror that distorted Boyle's preference for matters of fact recorded piecemeal, would he have wished to be called a natural philosopher? To withdraw the description altogether would seem incongruous given that, as both philosopher and theologian, he tackled Descartes's exclusion of final causes from natural philosophy, argued strenuously that the usefulness of natural philosophy included spiritual edification as well as practical utility, and analysed the grounds for asserting the superiority of a 'mechanical philosophy' of nature over others.

In Susan McMahon's essay we see how natural history, as practised and developed by John Ray, was itself transformed into an orderly discipline. The science of botany was no longer constituted by a miscellany of facts but elevated by its taxonomic goals and principles. Her story firmly situates natural history as a 'contingent social and cultural enterprise of time and place'. Ironically, the attacks on Ray's methodology from the Scotsman Patrick Blair served only to consolidate Ray's authority in England.

Given the popularity of Ray's *Wisdom of God Manifested in the Works of Creation*, it is not difficult to see how taxonomic practices could themselves unite a natural history to a natural theology. As long as the order imposed by the taxonomist could be said to reflect the order imposed by the creator, there was a happy congruence and aesthetic satisfaction for the botanist. Later in the eighteenth century, Linnaeus would make the connection in a slightly different way. If God had made the world a place of order and beauty and human beings alone had the capacity to study and appreciate that order, then the pursuit of the sciences was nothing short of a religious duty.[5]

McMahon's essay illustrates the fruitfulness of a historiographical principle that she makes explicit when speaking of Blair's belated recognition that 'the negotiation of differing intellectual opinions or diverging scholarly judgments depended not only upon the commitments of the individual members of the [botanical] community, but also upon the dedication of those members to perpetuate a particular communal identity and to promote a specific set of disciplinary standards'. Her call for the study of communities in their local contexts reinforces a well-established trend and is exemplified in Michael Honeybone's discussion of a truly localised society: the Spalding Gentlemen's Society. Articulating the values of that society, Honeybone stresses the virtue of curiosity and an understanding of the utility of science that emphasised aesthetic appreciation. Localisation did not have to mean cultural isolation. There was a link with Linnaeus in that members' desire for an ordered science of flora and fauna was partly satisfied by his new system, championed by an SGS member John Hill. The association of natural history with a delight in natural theology is central to Honeybone's account. A 'fundamental acceptance of natural theology' is said to have provided motivation for the society's members, among whom almost a quarter were Anglican clerics.

The problems that arose in classifying minerals are given detailed treatment in Matthew Eddy's searching study of the vocabularies employed by the Scottish mineralogists of the late eighteenth century. The influence of the Linnaean model is acknowledged, despite the problems it posed for Edinburgh's chemically trained physicians. Interestingly, as Eddy observes, it was also from Sweden that more finely tuned chemical vocabularies arrived to assist the chemists of the Edinburgh Medical School. Eddy's analysis provides fresh insight into the complicated relations between medicine, botany, chemistry, mineralogy and the nascent science of geology in this period: 'the professors of the Medical School were transferring chemical language into geology via mineralogy'. One consequence was that in that particular context geology did not yet pose a threat to a biblically sanctioned chronology. The formation of stones and minerals was not understood to require additional time. As Eddy puts it, 'scientific experimentation and theological beliefs could exist side by side within

---

[5] J. Brooke, *Science and Religion: Some Historical perspectives*, Cambridge, 1991, 192–

the same epistemological framework'. Even in cosmopolitan Edinburgh, where David Hume had exposed the defects of natural theology, we are still some way from the Victorian *conflict* between science and religion.

What transformations in the period from Newton to Darwin are illuminated by the various contributors? Transformations of at least three kinds receive attention: changes within a science, the cultural consequences of cheap brands of popular literature, and transformation in the orientation of individual scientists such as Darwin himself. A fundamental change in the character of the life sciences – from the taxonomic enterprise of natural history to the organicist biology of Bichat, Lamarck and Cuvier – underlies Richard Somerset's comparative study of the discipline whose business it is to document and explain transformation: history itself. His instructive comparison between Michelet and Carlyle is all the more vivid because it focuses on their respective reactions to a major political transformation – the revolution and its aftermath in France. This was a transformation Michelet could celebrate because it released the spirit of liberty; for Carlyle it was a punishment for moral abuse. As has often been observed, a conservative backlash against the more terrible aspects of the revolution helped to keep alive in Britain an association between science and religious propriety that was in some measure necessary to avert suspicion. In his 'Signs of the Times' (1829), in the very passage cited by Somerset, Carlyle would observe that 'This is not a Religious age. … Our true Deity is Mechanism'. And yet that very mechanism could still be the basis of a natural theology in which a more traditional deity was inferred from the excellence of his machinery – the very argument on which Paley had lavished so much care. Somerset's contrast between Michelet and Carlyle is subtle and revealing: 'Michelet must put history into the hands of Man if his Liberty is to have any meaning, and Carlyle must take history out of the hands of Man if its morally-conceived normative curve is to have any credibility as an eternal given'. In Britain in 1830 there were still eternal givens to be found, even if the historical *natural* sciences, paleontology for example, were beginning to chip at the biblical cadences.

The threat to fidelity charted by Aileen Fyfe arose from a different transformation, in which cheap secular literature increasingly left its mark on an expanding reading public. Twenty-five years after Carlyle's signs of the times, Cardinal Nicholas Wiseman would be complaining of the venomous, reptilean character of cheap publications. Fyfe's analysis shows how widespread was the concern in the 1850s that religious unbelief was increasing, deriving in part from damaging science but transmitted by affordable periodicals often silent on religious matters or, worse, anti-Christian in tone. Reference to the tone, and not merely the content, of such literature shows valuable insight, helping us to understand the sentiment, expressed by one of her subjects, that a secular work 'is no neutral, but an enemy cruising under a neutral flag'. As David Kohn has observed of Darwin's reticence, silence on religious matters can itself be one mode of secularisation.[6] Fyfe's essay furnishes ample evidence that a former liaison between science and fidelity was losing its potency as populist creeds became more materialistic and naturalistic. And this before Darwin's *Origin of Species* hit the headlines. Thomas Pearson's verdict that infidelity was 'coming in like a flood' was announced in the previous decade.[7]

Darwin's own religious odyssey repeatedly attracts attention as a case-study in the third kind of transformation identified above – the changing beliefs of an individual deeply absorbed by scientific and metaphysical issues. In Darwin there is a personification of the major cultural transition we are considering. In simple terms there was a movement from a Christianity sufficiently secure for him to study for the ministry, to a deism in which he was still willing to speak of a creator who had impressed laws on nature, to an agnosticism late in life when scientific insights into the order of nature no longer required reference to purpose. As Momme von Sydow observes, there was irony in the transformation. Darwin, of course, saw it very plainly: 'Considering how fiercely I have been attacked by the orthodox it seems ludicrous that I once intended to be a clergyman'.[8] For von Sydow the irony runs deeper because the science that led to Darwin's agnosticism had originally been grounded in the very natural theology he is routinely supposed to have destroyed. It is helpful to be reminded of a historiographical tradition in which continuities and homologies between Paley and Darwin have been recognised. Darwin's debt to Paley can be exaggerated if abstracted from other cultural influences; but von Sydow is correct in observing that there was more to Paley's philosophy of nature than a preoccupation with contrivances. Laws of nature featured prominently in Paley's vocabulary as they did in what David Kohn has called the *reformed* natural theology of John Herschel and Darwin.[9] This means that the ironic story is not confined to Darwin's biography since the basis of a scientific naturalism was already present in Paley.

The challenge, met by Darwin's biographers in different ways, has been how much weight to give the scientific roots of his agnosticism when we know that he was deeply affected by considerations of a quite different kind. Recent scholarship has tended to focus on the more contingent engines of unbelief: his doubts about revelation, his moral aversion to the preaching of eternal damnation for religious deviants, his sensitivity to the problem of pain and suffering, the existential crisis produced by the loss of his daughter Annie in 1851 and an ever hardening conviction that contingency itself, rather than providential intent, lies behind the accidents of daily life. By contrast

---

225.

6 D. Kohn, 'Darwin's Ambiguity: the Secularization of biological meaning', *BJHS* (1989), **22**, 215–39.

7 See also J. Secord, *Victorian Sensation: the Extraordinary publication, reception, and secret authorship of vestiges of the natural history of creation*, Chicago, 2000, 514.

8 C. Darwin, *The Autobiography of Charles Darwin* (ed. N. Barlow), London, 1958, 57.

von Sydow seeks to redress the balance by reinstating the internal logic of Darwin's ideas, insisting on a 'self-undermining' dynamic as his theory of natural selection undermined the presuppositions of the world-view from which it had partly sprung. On this reading Darwin's science is restored as a 'main cause' for his loss of faith. Whether one is persuaded or not, the conclusion nicely captures the complexity of an iconic transformation: 'belief and science are neither disconnected nor simply opposed to each other, but interwoven, mutually influencing and undermining each other and indirectly even undermining themselves'.

The master-narrative, represented by Frank Turner's account of the mid-Victorian conflict between science and religion, might be reintroduced at this point because T.H. Huxley's well-known exchange with Bishop Samuel Wilberforce in 1860 has epitomised the suppression of the clerical amateur naturalist by the rigorous professional scientist. As usually told, the story of the bishop's impertinence and comeuppance is not short on invention and revisionist accounts have been in circulation for some time. It is, however, to Frank James that we have been indebted for the most thorough literature search and for the striking result that during the 20 years following the encounter there was barely a mention of the event. Though Huxley's alleged triumph later became one of the foundation myths of scientific professionalism (and naturalism), the blank aftermath has supported the view, commended by James, that the significance of the debate has been much exaggerated. His contribution to this volume has the additional merit of identifying the sensibilities within the British Association that led to the drawing of a veil over an unseemly display. Against the universal mythic quality of the legend, he argues for the crucial role of local Oxford politics in determining perceptions of Wilberforce's authority.

Having reached the post-Darwinian debates we might reflect on some qualifications to the standard view that makes Darwin largely responsible for the demise of natural theology. Turner's essay already argued for a shift of focus – from an exclusive preoccupation with the relations between ideas to the social processes involved in establishing 'science' as a profession. Having devoted years to his magisterial study of John Henry Newman, Turner has also been sensitive to the fact that Christian theologians, as well as professionalising scientists, could be distrustful of physico-theologies.[10] In his biography of Michael Faraday, Geoffrey Cantor rightly insisted on the distinction between a natural theology and a theology of nature.[11] To recognise the theological limitations of *inferences from nature* did not mean that one's theological presuppositions could not find expression in the framing of metaphysical and even

[9] Kohn (1989).

[10] F. Turner, *John Henry Newman: the Challenge to Evangelical religion*, New Haven and London, 2002, 332 and 584. For evangelical suspicions of natural theology see J. Topham, 'Beyond the "Common Context": the Production and reading of the *BridgewaterTreatises*', *Isis* (1998), **89**, 233–62.

empirical questions about the natural world. There were other ways than via a natural theology in which religious beliefs might shape and frame questions about the workings of nature and Cantor brought them to light when discussing, for example, Faraday's commitment to principles of conservation and interconvertibility in his work on the correlation of physical forces. Students of Faraday have often welcomed his aphorism that religion and (natural) philosophy were two distinct things – which if a widespread view in the 1830s and 1840s would constitute another qualification to narratives that make the Darwinian impact central to the dissociation of scientific effort from religious sensibility. In his contribution to this volume, however, Cantor rejects what he sees as a facile interpretation of Faraday's remark about the independence of the two domains. Once placed in its immediate context, that of an embarrassed Faraday seeking to fend off the importunate attentions of an aristocratic and self-styled high-priestess of nature, with her heady combination of Swedenborgian, Roman Catholic and Rosecrucian spiritualities, his simple response, both tactful and tactical, becomes the more comprehensible. This need not suggest that he did not mean what he said, but the remark ceases to be a platform on which to erect a superstructure of interpretation.

That different contexts can produce different meanings is also a feature of Thomas Dixon's essay, in which the social sciences are given a high profile in the agency of secularisation. He notes that whereas today it is the theologians who are seeking to rescue genuinely altruistic behaviour from the sociobiologists who would explain it away, in Victorian contexts it was rather the converse as Comte's ascription of an innate benevolence to human nature was deployed against the more pessimistic doctrines of the Christian tradition. The fact that Comte drew on the brain sciences to substantiate his doctrine gives further depth to the perspective Dixon favours – one that is explicitly critical of Darwin-centred accounts. Attempts made by Comte and others to develop a science of ethics and of society constituted 'as widely discussed a threat to religious orthodoxy in Britain for most of the Victorian period as any geological or evolutionary theories, even those associated with the name of Charles Darwin'. Dixon also looks back to Frank Turner's early work, *Between Science and Religion* (1874), endorsing the value of studying those we might call the 'betweeners', who looked for middle ground between the often evangelical Christianity they had rejected and the doctrinaire scientific naturalism which seemed an inadequate substitute. Comte's concept of altruism had special appeal for those seeking to 'dissociate unbelief from decadence, immorality and revolutionary politics'. The evidently political character of their efforts leads Dixon to conclude his essay with the still necessary reminder that the abstractions 'science' and 'religion' are insufficiently refined tools for the examination of political realities.

One reason for that insufficiency is that what is to be included in the definition of 'science' can itself be contentious, just as it had been with natural philosophy. In Bill Brock's sensitive account of the conflict between William Crookes and

William Carpenter it becomes clear that the former was willing to include within the boundaries of empirical science 'the intangible, the odd, quirky and disagreeable'. Carpenter, deeply critical of his dallying with dubious mediums, was not. Through his investigation of the radiometer, Crookes was willing to see a connection between a material, scientific reality and a 'mysterious and religious unknown'. Not so Carpenter who believed that Crookes' conversion to spiritualism made him a biased judge of the evidence adduced in its support. The fact that two *professional* scientists could be so embattled on the issue of a special psychic force brings us back to that aspect of the master-narrative which directly associates the rise of scientific professionalism with the demise of natural theology. Brock's essay points up the divisions between professionals. The complication that professional standards within physics had not required the rejection of references to design and purpose, even if in evolutionary biology they were beginning to do so, has already been noted in connection with the Christian theism of Faraday, Maxwell, and William Thomson. The fact that it is Crookes the physicist, not Carpenter the physiologist, who has the more capacious definition of science embellishes this picture of tensions between what had now become specialised sciences and not merely different aspects of a broader natural philosophy. As Brock points out, Crookes stood on his dignity as a physicist. Not only that but the experimental physicists of the Physical Society, particularly George Carey Foster (professor of Physics at University College, London) were willing to rebut at least some of Carpenter's charges. As Brock observes, there is an irony in his story because the inspiration for some of Crookes' most productive work ultimately stemmed from his concept of a radiant matter that he believed might link our world to others.

The question whether arguments from nature to nature's god could survive the Darwinian impact has been answered in the affirmative for a different reason. Although Darwin undoubtedly contributed to a transformation in which any vestige of theological language was finally purged from technical scientific texts, modest and malleable inferences to a transcendent wisdom continued to feature in works of scientific popularisation. And they still do, with physicists prominent among those with seemingly more than a passing interest in the mind of God. In Peter Bowler's final chapter we see evidence for that correlation early in the twentieth century, particularly where doubts about the sufficiency of natural selection encouraged teleological readings of the evolutionary process. But Bowler's account of J. Arthur Thomson's career as a science populariser also brings this book to a poignant close as a widening gulf between the scientific specialist and the populariser finally meant that, by very virtue of being a populariser, one ran the risk of ostracism by the experts and of being overtaken by the new knowledge they produced.

In a recent discussion of the displacement of natural philosophy by natural science, Ronald Numbers reports that 'natural philosophers had often expressed a preference for natural causes, but few, if any, had ruled out appeals to God'. In contrast, 'virtually all scientists ... whether Christians or non-Christians, came by the later nineteenth century to agree that God talk lay beyond the boundaries of science'.[12] This is the aspect of the transition from natural philosophy to natural science that I have been addressing. I have tried to avoid the mistake of conflating natural theology with natural philosophy, while recognising that discussion of the existence and attributes of God, as by Newton, had often been deemed to *belong* to natural philosophy. Two recent studies remind us that qualifications are in order at the beginning as well as the end of the story, especially if we are tempted to think that this belonging together was typically a happy or stable union. In an important study of science as a calling, Mordechai Feingold has exposed the problems faced by seventeenth-century clerics who wished to devote more time than was considered seemly to their pursuit of natural knowledge. They experienced anxiety, had to contend with perceptions of incongruity, commented explicitly on the difficulty of maintaining two loyalties and devised their natural theologies partly in self-justification.[13] By the nineteenth century their difficulty was to be compounded by the impossibly escalating demands of a loyalty to scientific erudition. Secondly, as Peter Harrison has reminded us, early in the seventeenth century it was far from clear that an enquiring mind was a virtuous mind when the enquiry was directed towards the hidden operations of nature. With Francis Bacon's re-drafting of the doctrine of the fall in mind, Harrison asserts that 'the gradual transformation of curiosity from vice to virtue is an integral part of a larger story in which moral sensibilities delimit the sphere of legitimate knowledge and determine which natural objects are worthy of serious attention'.[14] This emphasis on moral sensibility leads to a final contrast between natural philosophy and the science to which it would eventually lead: 'While it might be possible to discern in the activities of Early Modern natural philosophers something that appears to answer to a "scientific method", it would be a mistake to conclude that the historical actors themselves regarded the legitimacy of their enterprise as dependent solely upon that method'. It is more accurate to say they were engaged in 'a morally informed "natural philosophy" and not a putatively objective "science".'[15] Whether that presumed objectivity, when it became the clarion call of science and had intrinsic moral value invested in it, necessarily extinguished older spiritual values is a question David Knight has himself addressed with his customary wisdom and vivacity.[16]

---

[11] G. Cantor, *Michael Faraday: Sandemanian and scientist*, Basingstoke, 1991, 162.
[12] R. Numbers, 'Science without God: Natural laws and Christian beliefs', in D. Lindberg and R. Numbers, eds, *When Science and Christianity Meet*, Chicago, 2003, 265–85, 272.
[13] M. Feingold, 'Science as a Calling? The Early Modern Dilemma', *Science in Context* (2002), **15**, 79–119.
[14] P. Harrison, 'Curiosity, Forbidden Knowledge, and the Reformation of Natural Philosophy in Early Modern England', *Isis* (2001), **92**, 265–90, 289.
[15] *Ibid.*, 290.

Index